Lecture Notes in Networks and Systems

Volume 880

The series “Lecture Notes in Networks and Systems” publishes the latest developments in Networks and Systems—quickly, informally and with high quality. Original research reported in proceedings and post-proceedings represents the core of LNNS.

Volumes published in LNNS embrace all aspects and subfields of, as well as new challenges in, Networks and Systems.

The series contains proceedings and edited volumes in systems and networks, spanning the areas of Cyber-Physical Systems, Autonomous Systems, Sensor Networks, Control Systems, Energy Systems, Automotive Systems, Biological Systems, Vehicular Networking and Connected Vehicles, Aerospace Systems, Automation, Manufacturing, Smart Grids, Nonlinear Systems, Power Systems, Robotics, Social Systems, Economic Systems and other. Of particular value to both the contributors and the readership are the short publication timeframe and the world-wide distribution and exposure which enable both a wide and rapid dissemination of research output.

The series covers the theory, applications, and perspectives on the state of the art and future developments relevant to systems and networks, decision making, control, complex processes and related areas, as embedded in the fields of interdisciplinary and applied sciences, engineering, computer science, physics, economics, social, and life sciences, as well as the paradigms and methodologies behind them.

Indexed by SCOPUS, EI Compendex, INSPEC, WTI Frankfurt eG, zbMATH, SCImago.

All books published in the series are submitted for consideration in Web of Science.

For proposals from Asia please contact Aninda Bose (aninda.bose@springer.com).

Domenico Marino · Melchiorre Alberto Monaca
Editors

Generative Artificial Intelligence and Fifth Industrial Revolution

Springer

Editors
Domenico Marino
Mediterranea University of Reggio Calabria
Reggio Calabria, Italy

Melchiorre Alberto Monaca
Mediterranea University of Reggio Calabria
Reggio Calabria, Italy

ISSN 2367-3370 ISSN 2367-3389 (electronic)
Lecture Notes in Networks and Systems
ISBN 978-3-031-73879-1 ISBN 978-3-031-73880-7 (eBook)
https://doi.org/10.1007/978-3-031-73880-7

This Springer imprint is published by the registered company Springer Nature Switzerland AG
The registered company address is: Gewerbestrasse 11, 6330 Cham, Switzerland

Contents

Introduction

Domenico Marino and Melchiorre Alberto Monaca

On the 1th and 2th of December 2023 International Seminar ***Generative artificial intelligence and fifth industrial revolution*** (AIMED2023) was held in Reggio Calabria, a welcoming Mediterranean city in Italy.

The approach that guided the design and development of the Aimed Seminar tried not to emphasize the positive aspects and not to hide the AI risks. As in any social phenomenon, the governance of processes and the definition of a system of rules and policies are the aspects that can create advantages or disadvantages and above all they can determine who are the beneficiaries of the advantages and who must bear the disadvantages.

The involvement of professors and researchers from Italian and Foreign Universities allowed us to develop and enrich scientific knowledge on Artificial Intelligence applied to the following fields: economics, finance, law, planning, operational techniques, management, production, human resources, business organization.

The level of some of the works presented and discussed in the forum on teaching and research experiences which AIMED2023 makes up led us to consider the possibility of compiling them in a document so that they could be published by world-renowned publishing house.

In this digital era, artificial intelligence (AI) is emerging as a catalyst for transformation across numerous fields, ranging from, economics, environment, finance to healthcare, AI's integration into daily operations and strategic planning presents a pivotal shift towards data-driven decision-making and automation [1–3]. Each chapter in this volume addresses a unique aspect of AI, from theoretical frameworks and technological advancements to practical applications and ethical considerations.

D. Marino (✉) · M. A. Monaca
Mediterranea University of Reggio Calabria, Reggio Calabria, Italy
e-mail: dmarino@unirc.it

M. A. Monaca
e-mail: melchiorre.monaca@unirc.it

D. Marino and M. Monaca (eds.), *Generative Artificial Intelligence and Fifth Industrial Revolution*, Lecture Notes in Networks and Systems 880,
https://doi.org/10.1007/978-3-031-73880-7_1

This volume not only highlights the advancements and applications of AI but also addresses the critical challenges of bias, privacy, and ethical implications associated with AI deployment. Through a multidisciplinary approach, it aims to provide readers with a nuanced understanding of AI's role in modern society and its potential to address some of the most pressing challenges of our times. As we stand on the brink of technological revolutions, this volume serves as a guide and a critical examination of the potential pathways AI might forge in the future. It is an essential read for academics, industry professionals, policymakers, and anyone interested in the profound changes AI is poised to bring. This is a multi-author book, but not a collection of essays. In fact, although signed by different authors, all the chapters of the book follow a line of development which is traced in the first part of the book and deepen the various aspects in logical order.

The various parts of the book will explore the most important features of AI and analyse the implications of AI in Economics, Law, Policies, Smart Citizens and Territorial Aspects.

References

1. Marino D, Gil Lafuente J, Tebala D (2023) Innovations and development of artificial intelligence in Europe: some empirical evidences. Eur J Manage Bus Econ 32(5):620–636. https://doi.org/10.1108/EJMBE-03-2023-0085
2. Tebala D, Marino D, Bianchino A (2023) A provincial indicator to describe potential artificial intelligence. In: Marino D, Monaca M (eds.) Artificial intelligence and economics: the key to the future. Lecture notes in networks and systems, vol 523. Springer, Cham. https://doi.org/10.1007/978-3-031-14605-3_1
3. Marino D, Cananzi D, Aragona F (2024) Artificial intelligence and ethics challenges of technologies. In: Ethics and artificial intelligence. Studies in systems, decision and control, vol 519. Springer, Cham. https://doi.org/10.1007/978-3-031-50902-5_3

A Regional Indicator to Describe Potential Artificial Intelligence in Agriculture

Domenico Tebala and Domenico Marino

Abstract Artificial intelligence (hereafter AI) is synonymous with innovation and ICT. Politecnico di Milano provides this definition of AI: "Artificial Intelligence is the branch of computer science that studies the development of hardware and software systems endowed with capabilities typical of human beings and capable of autonomously pursuing a defined purpose by taking decisions that, until then, were usually entrusted to humans". Banking, finance and insurance, automotive, energy and logistics are the sectors most adopting artificial intelligence projects. What about agricultural companies? For the purpose of a better understanding of the phenomenon to date, this study analyzes the potential use of Artificial Intelligence (AI) in Italian agricultural companies through a composite index (using the Equitable and Sustainable Welfare methodology) that measures the territorial gaps and their predisposition or not to the use of artificial intelligence which is an essential prerequisite for achieving intelligent growth.

For this purpose, the present contribution unfolds with the following structure:

- description of the theoretical reference framework and of the indicators used concerning innovations and Information and Communications Technology (ICT) of agriculture companies";
- description of the methodology for the construction of the composite indicator;
- description of the results, also through a geo-referenced map of regions potentially prone to the use of artificial intelligence in agriculture;
- conclusions.

Keywords Artificial intelligence · Index · Regions · Agriculture

D. Tebala (✉)
National Institute of Statistics, 88100 Catanzaro, Italy
e-mail: tebala@istat.it

D. Marino
Mediterranea University of Reggio Calabria, 89100 Reggio Calabria, Italy

D. Marino and M. Monaca (eds.), *Generative Artificial Intelligence and Fifth Industrial Revolution*, Lecture Notes in Networks and Systems 880,
https://doi.org/10.1007/978-3-031-73880-7_2

1 Introduction

Artificial intelligence (hereafter AI) is synonymous with innovation and ICT. Politecnico di Milano provides this definition of AI: "Artificial Intelligence is the branch of computer science that studies the development of hardware and software systems endowed with capabilities typical of human beings and capable of autonomously pursuing a defined purpose by taking decisions that, until then, were usually entrusted to humans. Banking, finance and insurance, automotive, energy and logistics are the sectors most adopting artificial intelligence projects. What about agricultural companies? For the purpose of a better understanding of the phenomenon to date, this study analyzes the potential use of Artificial Intelligence (AI) in Italian agricultural companies through a composite index (using the Equitable and Sustainable Welfare methodology) that measures the territorial gaps and their predisposition or not to the use of artificial intelligence which is an essential prerequisite for achieving intelligent growth [1–3].

For this purpose, the present contribution unfolds with the following structure:

- description of the theoretical reference framework and of the indicators used concerning innovations and ICT of agriculture companies";
- description of the methodology for the construction of the composite indicator; in particular, the robustness of the method identified will be assessed through the "influence analysis", which makes it possible to verify if and with how much intensity the rankings of the composite indices change following the elimination of an elementary indicator from the starting set and therefore the indicators that "weigh" more (COMIC software—COMposite Indices Creator);
- description of the results, also through a geo-referenced map of regions potentially prone to the use of artificial intelligence in agriculture;
- Conclusions

1.1 Process to Calculate the Indicator

1.1.1 Description of the Theoretical Framework

The approach used involves the construction of macro areas (pillars) by aggregating elementary indicators. Both pillars and elementary indicators have been considered non-replaceable. The pillars and indicators used, all with positive polarity, are [4] Table 1.

Table 1 Macro areas and indicators

Macro areas	Indicators
Innovation	% Innovative companies out of total agricultural companies (Var1) % Innovative companies for mechanization out of total innovative agricultural companies (Var2) % Innovative companies for planting and sowing on the total innovative agricultural companies (Var3) % Innovative irrigation companies out of total innovative agricultural companies (Var4) % Innovative companies for soil processing out of total innovative agricultural companies (Var5)
ICT	% of computerized companies out of total agricultural companies (Var6) % of computerized companies for accounting on the total computerized agricultural companies (Var7) % of computerized companies for crop management out of total computerized agricultural companies (Var8) % of computerized companies for livestock management out of total computerized agricultural companies (Var9) % of computerized companies for management of related activities out of total computerized agricultural companies (Var10)

1.2 Methodology

The matrix relating to data on Italian companies was divided into four progressive steps:

(a) Selection of a set of basic indicators on the basis of an ad hoc evaluation model hinging upon the existence of quality requirements;
(b) Further selection aimed at balancing the set of indicators within the theoretical framework of the structure. Outcome indicators are impact indicators as the ultimate result of an action as a result of a stakeholder activity or process;
(c) Calculation of synthetic indices (pillars), by making use of the methodology proved more appropriate to obtain usable analytical information on the agricultural potential artificial intelligence of the Italian regions;
(d) Processing of a final synthetic index as a rapid empirical reference concerning the degree of potential agricultural Artificial Intelligence of Italian regions.

Missing values were attributed via the *hot-deck* imputation and, where not possible, with Italy's average value.

The choice of the synthesis method is based on the assumption of a formative measurement model, in which it is believed that the elementary indicators are not replaceable, which is to say, cannot compensate each other.

The exploratory analysis of input data was performed by calculating the mean, average standard deviation and frequency, as well as correlation matrix and principal component analysis. Since this is a non-compensatory approach, the simple aggregation of elementary indicators was carried out using the correct arithmetic average with a penalty proportional to the "horizontal" variability.

Normalization of primary indicators took place by conversion into relative indexes compared to the variation range (min–max).

Attribution of weights to each elementary indicator has followed a subjective approach, opting for the same weight for each of them. Since, in some cases, the elementary indicators showed different polarity, it was necessary to reverse the sign of negative polarities by linear transformation.

For the synthetic indicator calculation, we used the *Adjusted Mazziotta-Pareto Index* (AMPI) [5], which is used for the min–max standardization of elementary indicators and aggregate with the mathematical average penalized by the "horizontal" variability of the indicators themselves. In practice, the compensatory effect of the arithmetic mean (average effect) is corrected by adding a factor to the average (penalty coefficient) which depends on the variability of the normalized values of each unit (called horizontal variability) or by the variability of the indicators compared to the values of reference used for the normalization.

The synthetic index of the i-th unit, which varies between 70 and 130, is obtained by applying, with negative penalty, the correct version of the penalty method for variation coefficient (AMPI $\pm$), where:

$$AMPI_i = Mri - Sricvi \tag{1}$$

where Mri e Sri are, respectively, the arithmetic mean and the standard deviation of the normalized values of the indicators of the i unit, and $cvi = Sri/Mri$ is the coefficient of variation of the normalized values of the indicators of the i unit.

The correction factor is a direct function of the variation coefficient of the normalized values of the indicators for each unit and, having the same arithmetic mean, it is possible to penalize units that have an increased imbalance between the indicators, pushing down the index value (the lower the index value, the lower the level of potential regional agricultural AI).

Both the variance and the standard deviation are absolute indices of variability and are therefore affected of the unit of measurement and of the order of magnitude of the data. These indices do not allow the degree of variability to be compared of phenomena that have different units of measurement and of characters which, even if measured with the same unit of measurement, refer to very different collectives (therefore having equally different average values) The variation coefficient, as a relativized index, allows you to compare the variability of two or more distributions.

This method satisfies all requirements for the wellbeing synthesis [6]:

- Spatial and temporal comparison
- Irreplaceability of elementary indicators
- Simplicity and transparency of computation
- Immediate use and interpretation of the obtained results
- Strength of the obtained results.

An influence analysis was also performed to assess the robustness of the method and to verify if and with which intensity the composite index rankings change following elimination from the starting set of a primary indicator. This process has also permitted us to analyze the most significant indicators.

The analysis was conducted using the *COMIC* (Composite Indices Creator) software, developed by ISTAT. The software allows calculating synthetic indices and building rankings, as well as easily comparing different synthesis methods to select the most suitable among them, and write an effective report based upon results.

2 Description of the Results

Table 2 reveal a good variability while Table 3 show significant correlations between the analyzed indicators of the macro areas. In particularly, there are significant direct correlations between % Innovative companies out of total agricultural companies (Var1) and % Innovative companies for soil processing out of total innovative agricultural companies (Var5), % of computerized companies out of total agricultural companies (Var6) and % of computerized companies for accounting on the total computerized agricultural companies (Var7); between % Innovative companies for soil processing out of total innovative agricultural companies (Var5) and % of computerized companies out of total agricultural companies (Var6) and % of computerized companies for livestock management out of total computerized agricultural companies (Var9).

Instead there are indirect correlations between % of computerized companies for crop management out of total computerized agricultural companies (Var8) and Innovative companies out of total agricultural companies (Var1) and % of computerized companies out of total agricultural companies (Var6); between % of computerized companies for livestock management out of total computerized agricultural companies (Var9) and % Innovative companies for planting and sowing on the total innovative agricultural companies (Var3), % Innovative irrigation companies out of total innovative agricultural companies (Var4) and % of computerized companies for crop management out of total computerized agricultural companies (Var8).

The influence analysis describes the indicators that most influence the composition of rosters in potential agriculture IA of regionss. In analyzing Table 4 we can see that the most important indicator concerns % Innovative irrigation companies out of total innovative agricultural companies (Var4) (mean = 1.500, σ = 1.466), %

Table 2 Mean, σ and frequency indicators

	Var1	Var2	Var3	Var4	Var5	Var6	Var7	Var8	Var9	Var10
Mean	13.61	6.06	9.925	4.29	4.8	19.83	72.18	28.375	17.125	15.985
σ	8.55	1.652	3.341	2.231	2.061	**13.333**	4.378	5.674	7.48	4.224
Frequency	20	20	20	20	20	20	20	20	20	20

Table 3 Correlation matrix of indicators

Indicators	Var1	Var2	Var3	Var4	Var5	Var6	Var7	Var8	Var9	Var10
Var1	1.000									
Var2	− 0.025	1.000								
Var3	0.183	− 0.266	1.000							
Var4	− 0.298	− 0.326	0.409	1.000						
Var5	**0.527**	0.097	− 0.250	− **0.520**	1.000					
Var6	**0.987**	0.043	0.120	− 0.347	**0.602**	1.000				
Var7	**0.500**	0.019	0.417	0.226	− 0.103	0.466	1.000			
Var8	− **0.573**	− 0.147	0.265	0.398	− 0.328	− **0.591**	− 0.406	1.000		
Var9	0.258	− 0.074	− **0.585**	− **0.589**	**0.728**	0.340	− 0.290	− **0.503**	1.000	
Var10	− 0.044	0.812	− 0.020	− 0.030	− 0.185	− 0.023	0.041	0.120	− 0.443	1.000

The values with high correlations have been highlighted

Table 4 Influence analysis: mean and σ of the shifts of the rankings by basic indicator removed of indicators

Indicators	Mean	σ
Var1	1.200	1.208
Var2	1.250	1.401
Var3	1.600	1.356
Var4	**1.500**	**1.466**
Var5	1.000	1.000
Var6	1.400	1.319
Var7	**2.100**	**1.513**
Var8	1.200	1.860
Var9	1.500	1.162
Var10	**1.500**	**1.658**
Mean	**1.425**	**1.394**
σ	**0.286**	**0.236**

The highest values of mean and standard deviation have been highlighted

of computerized companies for accounting on the total computerized agricultural companies (Var7) (mean = 2.100, σ = 1.513), and % of computerized companies for management of related activities out of total computerized agricultural companies (Var10) (mean = 1.500, σ = 1.658).

3 Discussion and Conclusions

The cartographic representation of the value of the final synthetic index does not fully reflect the usual dualistic scheme of the South/Central-North (Table 5 and Fig. 1).

In particular, the best performances are grouped in Friuli-Venezia Giulia, Lombardy and Umbria, but the region where artificial intelligence is potentially most widely used in agriculture is Trentino Alto Adige (Index 111.23), thanks mainly to the highest percentages of innovative (38%, Italy 11%) and computerized companies (55.6%, Italy 15.8%).

Molise occupies the last position in the ranking (Index 90.2) and follows the negative trend of the South from which the following stand out:

- Campania (10th place—Index 99.0) for the highest percentage of computerized companies for crop management (36.2%, Italy 28.2%),
- Calabria (12th place—Index 97.1) for the second highest percentage of companies that have made innovative investments in irrigation (8.7%, Italy 4.5%)
- Sicily (13th place—Index 95.8) for the third highest percentage of companies that have made innovative investments in irrigation (7.3%, Italy 4.5%)

Table 5 Synthetic Regional Index ranking of potential agriculture IA

Regions	Value	Rank
Trentino Alto Adige/Sudtirol	109.071	1
Lombardia	105.290	2
Friuli-Venezia Giulia	102.863	3
Umbria	101.438	4
Toscana	101.235	5
Veneto	100.659	6
Liguria	100.262	7
Emilia-Romagna	99.704	8
Piemonte	99.428	9
Campania	99.081	10
Lazio	97.722	11
Calabria	97.162	12
Sicilia	95.863	13
Sardegna	94.284	14
Abruzzo	94.180	15
Marche	93.889	16
Puglia	93.866	17
Valle d'Aosta/Vallée d'Aoste	92.638	18
Basilicata	92.464	19
Molise	90.214	20
Italy	100.000	

As part of one of the 6 priorities for 2019–2024, namely 'A Europe fit for the digital age', the European Commission has proposed a set of actions to promote excellence in AI and rules to ensure that the technology is reliable, ensuring the safety and fundamental rights of people and businesses, while strengthening investment and innovation in EU countries [7–9].

In this context, the synthetic index just described can be useful to give us an idea of the use, or otherwise, of artificial intelligence at a territorial level in the agricultural sector, but above all it can constitute a support for the decisions of European and Italian policy makers,

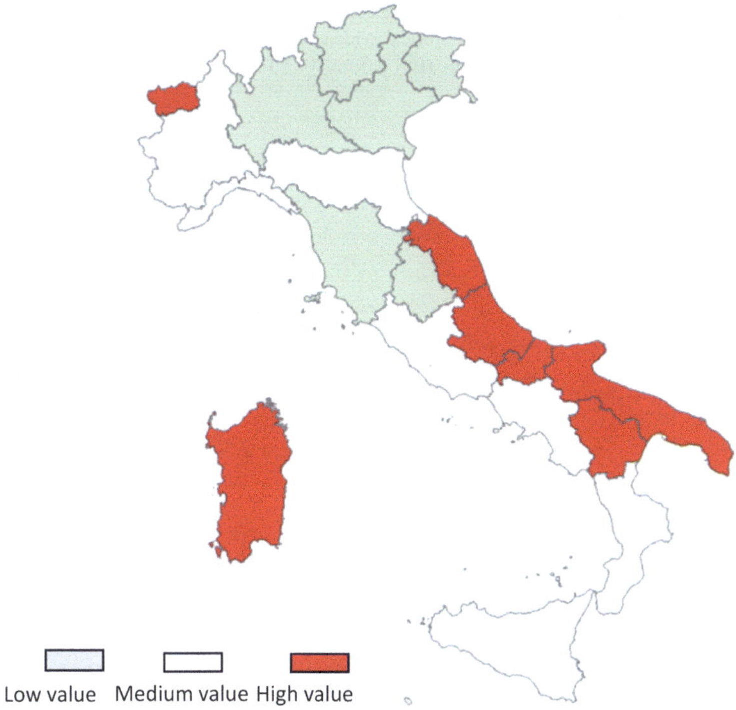

Fig. 1 Territorial distribution of the regional synthetic index of potential agriculture IA

References

1. Cuomo C, Marino D (2019) The local work plans (LWP) and territorial economic system (TES): assessment and evaluation. Studies in systems, decision and control, vol 180. pp 225–237, ISSN: 2198-4182. https://doi.org/10.1007/978-3-030-00677-8_19
2. Marino D, Tebala D (2021) Economy and development in Calabria: the weak development hypothesis. Rivista Internazionale di Scienze Sociali–anno CXXIX 2:187–220, ISSN: 0035-676X. https://doi.org/10.26350/000518_000068
3. Tebala D, Marino D, Bianchino A (2023) A provincial indicator to describe potential artificial intelligence. In: Marino D, Monaca M (eds) Artificial intelligence and economics: the key to the future. Lecture Notes in Networks and Systems, vol 523. Springer, Cham. https://doi.org/10.1007/978-3-031-14605-3_1
4. Istat: 7° Censimento generale dell'Agricoltura (2021). https://www.istat.it/it/censimenti/agricoltura/7-censimento-generale
5. Mazziotta M, Pareto A (2016) On a generalized non-compensatory composite index for measuring socio- economic phenomena. Soc Indic Res 127(3):983–1003
6. Istat: Bes 2015—Il Benessere Equo e Sostenibile in Italia (2015)
7. Marino D, Gil Lafuente J, Tebala D (2023) Innovations and development of artificial intelligence in Europe: some empirical evidences. Eur J Manage Bus Econ 32(5):620–636. https://doi.org/10.38191/iirr-jorr.20.022
8. Marino D, Tebala D (2022) Rural areas and well-being in EU countries + UK: a taxonomy and a cluster analysis. Sustainability 2022(14):15213. https://doi.org/10.3390/su142215213

9. Marino D, Timpano F, Gil Lafuente J (2023) The competition between legal and illegal firms in the market: theoretical models and empirical evidence. Eur Res Manag Bus Econ 29(3):100228. ISSN 2444-8834. https://doi.org/10.1016/j.iedeen.2023.100228
10. http://dati.istat.it

Ethics Effects on the Investment Decision Making Process

Fabio Piluso, Caterina Di Tommaso, and Danilo Antonio Ceraso

Abstract Investor's behavior is affected by multiples factors. In this paper, we investigate the factors affecting the decision-making process, the framing effect and the loss and risk aversion. Our findings suggest that female, younger, employed and graduated investors are more ethical investors. A good financial culture does not imply the overcoming of the cognitive bias. Not all categories of investors are affected by the attribute framing. Finally, we find empirical evidence of an aversion to loss bias. After a loss occurs, investors may decide to shift to an ethical fund for moral principles ("*ethics effect*") in addition to the isolation effect.

Keywords Sustainable finance · Corporate social responsibility · Socially responsible investments · Financial culture

1 Introduction

The decision-making process has been long studied by academics and practitioners. For decades, economics, psychology, sociology, political science scholars have tried to explain this process. The theories and models of neoclassical economics have embraced the assumption of the perfect rationality of individuals. Over time, this line of theoretical studies has proved far away from catching some economic hypothesis (for example asymmetric information) and therefore, these models have not been sufficient to justify possible irrational or sub-optimal decisions of individuals. Especially in complex decisions, people are unable to analyze the whole scenario and, very

F. Piluso (✉)
Economics of Financial Intermediaries, University of Calabria, Cosenza, Italy
e-mail: fabio.piluso@unical.it

C. Di Tommaso
Economics of Financial Intermediaries, University of Bari, Bari, Italy

D. A. Ceraso
Economics and Business, University of Calabria, Cosenza, Italy

D. Marino and M. Monaca (eds.), *Generative Artificial Intelligence and Fifth Industrial Revolution*, Lecture Notes in Networks and Systems 880,
https://doi.org/10.1007/978-3-031-73880-7_3

often, choose sub-optimal solutions. Economic models based on perfect rationality may find difficult to explain this type of economic behavior. Developments in cognitive and social psychology have played an important role in changing main economic theories. Since the 1950s, some economists have questioned the assumption of perfect rationality [28]. Psychology has provided a new set of variables that could impact on the decision-making process and be easily implemented within existing neoclassical models. This new line of studies is now called behavioral economics.

More recent neuroscience researches have also allowed us to better investigate the role of the brain in decision making processes. Neuroscience, for example, offers a scientific approach to psychology and has the potential to confirm or overlook psychological theories. Starting in the late 1990s, the economic branch called "neuro-economics" began its development in order to integrate the results of behavioral economics with the most recent results of neuroscience. The interconnections between all these branches of behavioral finance offer a wide range of tools to better explain the irrational and sub-optimal choices that are not tied to hypotheses of perfect rationality.

Over the past several years, there has been increasing interest in ethical issues [22]. During period of economic crisis or exogenous shocks (such as the recent pandemic of Sars-Covid 19), there has been a great process of switching between traditional mutual funds and funds falling into the ESG or SRI category (hereinafter "ethical funds"). According to data provided by Morningstar (2020),[1] in the first quarter of 2020 positive global net flows are estimated for 45.6 billion dollars, against net redemptions for 384.7 billion from traditional instruments. Geographically, Europe drives collection, absorbing 72.4% of the total flows, albeit lower in quantitative and percentage terms than in previous quarters,[2] followed by the United States. The same trend is recorded for Japan, China and South Korea. According to the latest data from the SRI Atlas of EticaNews, in Italy, the SRI assets under management are close to 132 billion euros, of which 110.2 billion attributable to funds and 21.47 billion to ETFs traded on the Italian Stock Exchange.

The decision to invest in an ethical fund appears to be related to psychological and behavioral motivations and not to the greater performance of ethical funds. The comparison between MSCI World index SRI and MSCI world index (Fig. 1) reveals that the SRI fund's performance is similar to the mutual fund's performance in the period 2007–2020.

Research on the implications for cognitive biases of ethical investment is warranted for several reasons. First, the SRI funds market is still small, notwithstanding the growth of new investments described above. Second, the moral principles that guide the decision-making process could lead to discover different cognitive biases, which complicates analysis of investor's decision-making process. Third, there remains room for a wide interpretation of what constitutes "ethics investment" given the lack of consistent global standards, which exposes ethics investors to moral

[1] Report *Global sustainable fund flows* di Morningstar (2020).

[2] In Europe, sustainable funds recorded positive deposits of $33.14 billion in the first three months of the year against outflows of 163 billion in the other funds.

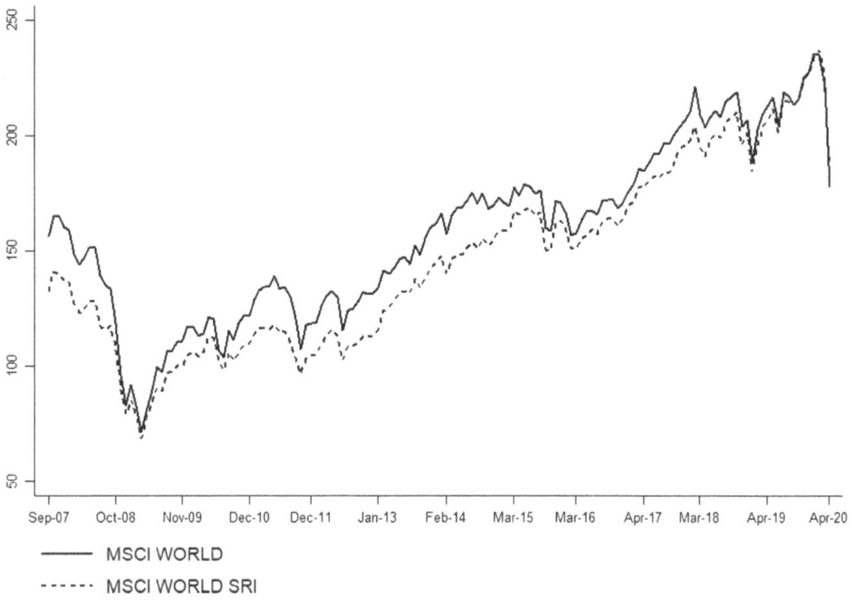

Fig. 1 MSCI world and MSCI SRI

risk. Researches investigating the existence of cognitive biases and their implications for investor's decision-making process were realized for mutual funds, it will be interesting to expand the research also for ethics funds. The aim of this paper is to study the ethical investment decision making process in period of economic and financial stability. In particular, we test the impact of different factors that has been shown to impact the decision-making process. In a first study, we test the factors impacting on the decision to invest in an ethical fund and the framing effect. In a second study, applying the prospect theory [16], we verify the certainty and isolation effect on the decision to invest in an ethical fund. To this purpose, we conducted a survey on 500 participants during the period July–September 2019. We report three key results. First, we find that the decision to invest in an ethical fund depends on a series of factors, such as gender, age, employment and education. This implies that some investors view ethical investment as a financial tool that reflects the moral principles in which they believe. Second, the attribute framing is verified for some type of investors. Thus, the positive frame appears to have provided a framework for ethical investments to be a credible mechanism to engage in ethical activities. Third, when we take account of loss and certain loss, the decision-making process appears to be strongly associated with an aversion to loss bias. Nevertheless, for some specific types of investors, the aversion to loss bias is greater because there is an additional effect that we call "ethics effect".

Our paper makes several contributions to the decision-making process literature. First, we contribute to the small literature on the determinants of factors impacting on the decision to invest in an ethical fund, most of which has focused on traditional mutual funds. In addition, most of these studies have focused on the determinants of the ethics investments by the corporate point of view, while our analysis focuses on the investors point of view. Second, we contribute to the literature studying the impact of the framing effect on the decision-making process. Our results show that the positive frame would lead to a change of mind in some type of investors. Third, we contribute to the still limited literature linking investment in ethics activities to loss aversion by showing that, after a loss occurs, the possibility to shift to an ethical fund is a significant driver of decision-making process.

The rest of the paper is organized as follows. In the next section we review the relevant literature and develop our hypotheses. Section 3 presents our method and sample. Our empirical results are discussed in Sects. 4 and 5 concludes.

2 Related Literature

The aim of this paper is to investigate the decision-making process in the ethical investment. First, we study the factors affecting the decision to invest in an ethical fund. Second, we focus on the framing effect and, finally, we investigate the concept of aversion to loss and aversion to certain loss.

Different authors have focused on the field of decision-making process that has been long studied but there are few papers that focus on the ethical investment decision making process despite the investors is increasing attention on ethical investments.

The most related literature to our paper is the literature studying the factors that may affect ethical decisions. In particular, there are several theoretical and empirical studies focusing on the different factors that may impact the decision-making process.

Rest [25] is the pioneer of the theoretical models. Rest [25] defines four-step model for individual ethical decision making and behavior: (1) recognize the moral issue, (2) judge a situation as being moral; (3) compare moral concerns and other concerns and (4) prioritize moral values over other values. The Jones [15]'s model extend the Rest [25]'s model by proposing an issue-contingent model. Jones [15] introduces the concept of moral intensity that is a construct that have different components: magnitude of consequences, social consensus, probability of effect, temporal immediacy, proximity, and concentration of effect.

The empirical contributions on this topic have studied many factors that may impact the ethical decision making process [6, 11, 17, 21, 24]. The determinants of the ethical decision making process can be divided into individual variables and situational variables [11, 24]. The individual variables include all those factors that are uniquely associated with the individual decision maker. The main individual factors are: nationality, gender, age, personality, attitudes, values, education, religion, employment, etc. The results about all these factors remain somewhat mixed and inconclusive.

As shown in Ford and Richardson [11] and later in the literature review of Loe et al. [21] and Craft [6], the most studied factor of the decision making process is the gender. The results about the impact of gender on the decision-making process are mixed. According to the Gender socialisation theory, most of the empirical studies [4, 7, 27] have found that women and men differ with regards to the ethics. In particular Ruegger and King [27] show that the females are more ethical than males in their perception of business ethical situations. Eweje and Brunton [10] note that, among 655 business students, the females are more ethically aware than their male counterparts. They find significant differences between males and females regarding ethical judgement. Dawson [7] specifies that the gender differences are significant only in case of relational issues and it becomes insignificant in case of non-relational situation. Finally, using managers as the research target, Chen et al. [4] show that the gender difference in ethics becomes more pronounced under the cultural dimensions of collectivism, humane orientation, performance orientation, and gender egalitarianism. Other studies, such as Robin and Babin [26] and McCabe et al. [23], reveal that no differences occur between men and women in their ethical perceptions.

Given the mixed findings and consistent with these views, we can formulate our first hypothesis in two different ways:

$H_{1(a)}$: *Females are more likely to invest in ethical funds than males.*

$H_{1(b)}$: *Ethical investments do not change with the gender.*

The literature studying the effect of age on the decision-making process show mixed results. Most of papers find that older people are more ethical than younger people [5, 9, 27]. Specifically, Ruegger and King [27] suggest that the aged people (40 plus years age) are the most ethical people. In analysing the impact of both individual ethics and organizational ethics on ethical intention, Elango et al. [9] indicate that younger managers were more likely to be influenced by organizational ethics than older managers, and that older managers were more likely to make ethical choices. In line with the previous findings, some studies have used the career stage as a proxy of the age [2]. They find that people in the early career stage act less ethical than in the later stages.

In contrast with the previous papers and as shown in the study of Ford and Richardson [11], there are studies that have shown a negative association between age and ethical decision making process. In particular, Browning and Zabriskie [3] is the first study that have demonstrated that the younger managers appear to be more ethical than the older manager. Furthermore, Ebrahimi et al. [8] find a negative association between age and ethical behavior. On the same direction, Lewis and Mackenzie [19] argue that younger and better educated investors prefer to put their money in ethical investments. In line with this literature, we can express our second hypothesis as follows:

$H_{2(a)}$: *Younger investors are more likely to invest in ethical funds than older investors.*

$H_{2(b)}$: *Older investors are more likely to invest in ethical funds than younger investors.*

As indicated in Ford and Richardson [11]; Lehnert et al. [17], the relationship between employment, education and ethical decision making is not significant.

In line with this finding, we state our third hypothesis as follows:

H_3: Employment and education do not impact the decision to invest in an ethical fund.

The second part of the paper focuses on testing the framing effect introduced by Tversky and Kahneman [30]. The framing effect is related to the decision making process and to the psychological principles that govern the perception of decision problems. Tversky and Kahneman [30] show that people may change the preferences when the same situation is described in different ways. The framing theory in behavioral finance refers to the way in which the information is presented. The investors may react to a particular information differently. In particular, according to the framing theory, it is possible to add different framing types. Specifically, in this study, the information is present in a positive and negative way (attribute framing). Some empirical evidence has demonstrated that the addition of an attribute framing can generate a framing bias in investment decision making process. For example, Agnew et al. [1], has shown that the negative framing can have a powerful effect on financial decision. In a more recent study, Lei and Zhang [18] have demonstrated that there is a framing bias in investment. They show a significant positive association between an uncertain or constraining media sentiment and socially responsible investment performance, as well as a negative relationship between a negative sentiment and socially responsible investment performance. Applying the framing theory to the ethical investment decision making process, we can specify our fourth hypothesis as follows:

H_4: Framing effect. The introduction of a positive frame would lead investors to change idea about the investment in ethical fud.

Finally, the purpose of this paper is to investigate the concept of "aversion to loss" and "aversion to certain loss" applied to the ethical investment decision making process. The literature studying the decision-making process under risk has been long linked to the expected utility theory. Nevertheless, in 1979, a paper of Kahneman and Tversky offers an alternative model, so called Prospect theory model. This model has shown that the aversion to loss and some other state of mind can impact the investment decision making process. Specifically, according to the Prospect theory, the concept of *"aversion to loss"* implies being more sensitive to losses than gains. The Prospect theory describes a behavioral bias by demonstrating that the investor become risk averse in two different cases: (a) prior return was above the target level and (b) previous loss. The model of Kahneman and Tversky [16] introduces the certainty and isolation effect. The certainty effect contributes to risk aversion in choices involving sure gains and to risk seeking in choices involving sure losses. This tendency is due to the individuals that underrate outcomes that are merely probable in comparison with outcomes that are obtained with certainty. By contrast, the isolation effect leads to inconsistent preferences when the same choice is presented in different forms. This is because investors generally do not consider components that are shared by all prospects under consideration. Therefore, applying the prospect theory to the ethical investment decision making process, we can specify our fifth hypothesis as follows:

Table 1 Description of the sample

Variable	Option	%
Gender	Men	48.4
	Women	51.6
Age	Under 21	6.2
	From 22 to 30	56.0
	Up to 21	37.8
Education	Graduated	56
	Not graduated	44
Status	Employed	67
	Not employed	33
Financial culture	Knowledge of financial products	60
	Not Knowledge of financial products	40
Yearly income	From 0 to 12,000 €	43
	From 12,000 to 32,000 €	41
	Up to 32,000 €	16
Risk appetite	Low	3
	Below average	14
	Moderate	47
	Above average	26
	High	10

H_5: Aversion to loss bias. Some state of mind such as aversion, loss aversion and mental accounting can affect an individual's decision-making process.

3 Method

3.1 Participants

Our sample consisted of people with different levels of education, age and gender to maximize the variance of the sample. The sample is composed by 500 participants. Participants were recruited by using the online section of google modules.[3] The data are collected during the period July–September 2019.[4] The composition of the sample is reported in Table 1.

[3] Website: https://docs.google.com.

[4] Stable period under the economic and financial point of view. Per effettuare l'indagine è stato utilizzato un questionario creato online nella sezione Moduli del sito https://docs.google.com.

The male respondents represent half of the sample. 56% of the sample are aged from 22 to 30 years old and the same percentage declares to be graduated. More than 50% of the participants are employed and have knowledge of financial products. 43% of the respondents have a yearly income from 0 to 12,000 € and 41% have a yearly income from 12,000 to 32,000 €. Finally, the sample show a risk appetite moderate in the half of the respondents. Only 10% of the participants can be defined as people with a high-risk appetite.

3.2 Design and Procedures

Our questionnaire is composed by 23 questions divided by personal information and risk tolerance assessment information. To test our hypothesis in relation to the determinants of the ethical investment decision making process, we ask the participants to provide information about gender, employment, age, education, income, financial culture. In the second part of the questionnaire, following the paper of Grable and Lytton [13], we ask the participants to answer 13 questions to evaluate the participants risk tolerance.

To measure the preferences to invest in an ethical fund, we insert two different questions. The first question asks participants to identify the preferences to invest in different financial solutions such as deposit savings into a bank's account, keeping savings at home, investing savings in an ethical fund and investing saving in an investment fund. In this context, 13.4% (67) of the participants indicate to choose the investment in an ethical fund whereas 86.6% (433) of the participants indicate to choose the other options. In order to test the framing effect, the second question adds the frame attribute to the previous question. In particular, it adds a positive frame for the option "investing saving in an ethical fund" and negative frames for the other options. Adding a positive frame leads the participant to change idea. Indeed, our questionnaire shows that 266 (53.2%) participants decide to invest in an ethical fund, whereas 234 (46.8%) indicate the other options. This implies that we have evidence about the framing effect on the ethical investment decision making process. This finding confirms our hypothesis H_4 and provides empirical evidence of a framing bias effect on investment decision making process.

Finally, in order to verify the effect of the concept of "aversion to loss" and "aversion to certain loss" [16] in the context of information processing for decision making, we insert two questions in which the participants had to deal with two situations presenting two possible choices. In the first situation, the respondents had to choose to exit from the investment after a loss of 10% and the possibility to remain on the financial position- half of the participants preferred to maintain the financial position and hope to recover the loss in a future rise of the financial markets. In the second situation the participants had to choose between the option to exit from the investment and the option to accept the transition to the ethical fund. In this scenario, the majority of the participants shown a strong aversion to certain loss and, as much, 72% preferred to accept the transition to the ethical fund.

4 Results

This section presents the results of two studies that were designed to verify the determinants of the decision to invest in an ethical fund, and the concept of "aversion to loss" and "aversion to certain loss". The first study is divided in three sub-studies aimed to understand how investors perceive ethical fund and to verify whether the determinants of the choice may change in relation to different specifications. Study 1A and 2A differ because of the formulation of the questions and options. In Study 3A, the aim is to test the *framing effect* and the factors able to make a change in the decision to invest in an ethical fund. Study B tests the factors that have a positive and negative impact on the "aversion to loss" and "aversion to certain loss". Study 1B and 2B differ on the formulation of the question and options and test the certainty effect. Finally, study 3B investigates the isolation effect.

4.1 Study A

To study the factors that have an impact on the decision to invest in an ethical fund, we estimate the following model:

$$\begin{aligned}\text{Investment decision} = {} & \beta_0 + \beta_1 * \text{Gender} + \beta_2 * \text{Age} + \beta_3 * \text{Status} \\ & + \beta_4 * \text{Education} + \beta_5 * \text{Income} + \beta_6 * \text{Financial culture} \\ & + \beta_7 * \text{Tolerance} + \varepsilon \qquad (1)\end{aligned}$$

In all the sub-studies, the dependent variable is a dichotomous variable and thus, to test this model, we apply a Probit model·[5] Among the independent variables, we test the impact of *gender*, a dummy variable taking the value one for male investors and zero for female investors. *Age* variable, as in Ruegger and King [27], is a scale variable. Investors were divided into 4 groups according to age: under 21 years, 22–30 years, 31–40 years, and 40 plus years. The *status* variable measures whether or not the investor is employed. *Education* considers the culture's degree of the investors. In particular, it is equals to one if the investor has a degree and zero otherwise. The income variable considers three yearly income brackets (from 0 to 12,000 €, from 12,000 to 32,000 € and ≥ 32,000 €). Finally, to account for the financial education and the risk tolerance we consider the variables: *Financial culture* and *Tolerance*. The financial culture is a binary variable equals to one if the investor declares that he/she knows the financial instruments and zero otherwise. To calculate the risk tolerance we use the methodology of [14], Linciano and Soccorso [20].

[5] The Logit model results support the conclusions from the Probit model and are available from the authors on request.

Table 2 Choices of the investors in Question 1

Options	*Question 1*: Imagine to have a monthly net income of 2000 € and to invest your own savings, which financial solution would you prefer (%)
Investing savings in an ethical fund	13.4
Keeping savings at home through the classic "money under the mattress"	5.4
Allocate savings in an investment fund chosen by you	41.6
Monthly deposit savings into a banks account	39.6

4.1.1 Study 1A

Study 1A aims at evaluating the factors that may affect the choice to invest in an ethical fund rather than in some other financial solutions. Table 2 reports the choices of the investors. The main financial solution preferred by the investors is the investment in a fund chosen by themselves. Furthermore, the investors prefer to save money by making a monthly deposit in their bank's account (39.6%) rather than invest in an ethical fund (13.4%). Only 5.4% of investors prefer to keep money at home.

To test the factors impacting on the decision to invest in an ethical fund, we implement the model in Eq. 1. In Study 1A the dependent variable, *investment decision,* measures the choice to invest or not in an ethical fund. The dependent variable is a dummy variable that takes the value one if the investor decides to invest in an ethical fund and zero otherwise.

The Probit model results of Study 1A are reported in column 1 of Table 5. The decision to invest in an ethical fund depends on the gender, age, employment, instruction, income and financial culture. Specifically, female, younger, employed and graduated investors are more likely to invest in an ethical fund. Furthermore, an increase on the yearly income increases the probability to invest in an ethical fund. Finally, investors that have a greater financial culture reduces the probability to invest in an ethical fund. Our results corroborate our hypothesis $H_{1(a)}$ and $H_{2(a)}$. Contrary to our hypothesis H_3, we find empirical evidence of a positive and significant impact of employment and education on the ethical investment decision making process.

4.1.2 Study 2A

Study 2A aimed to gather further support for the choose to invest in an ethical fund. In this study we have changed the question by:

a. specifying the reason why the investors have to save money for the future. This would take into account the mental accounting process [29],

Table 3 Choices of the investors in Question 2

Options	*Question 2*: Imagine to have a monthly net income of 2000 € and to invest your own savings to help your sons or your nephews that have to start the university in 10 years. With the goal of growing your funds and having to protect yourself from possible fraud on the market, which financial solution would you prefer (%)
Investing savings in an ethical fund (investment with a high social content and transparency)	53.2
Keeping savings at home through the classic "money under the mattress" (with all the risks associated with keeping cash at home)	2.8
Allocate savings in an investment fund chosen by you (with the possibility of obtaining returns if the choice proves correct, losses if the choice proves wrong)	16.6
Monthly deposit savings into a banks account (interest rates close to zero)	27.4

b. adding a positive frame for the option "*Invest in an ethical fund*" and negative frames for the other options [12].[6]

Table 3 reports the choices of the investors in Question 2. The investors that decide to invest in an ethical fund are 53.2%. 27.4% of investors prefer to have some liquidity in their banks account. 16.6% and 2.8% prefer to invest in a fund and to keep money at home, respectively.

As in the previous study, to investigate the determinants that may affect the choice to invest in an ethical fund, we run Eq. 1 by using a Probit model in which our dependent variable is equals to one if the investor declares to invest in an ethical funds and zero otherwise.

When the question adds the reason why to save money and positive/negative frames in the options, the determinants of the ethical decision-making process change

[6] Our question become: "Imagine to have a monthly net income of 2000 € and to invest your own savings to help your sons or your nephews that have to start the university in 10 years. With the goal of growing your funds and having to protect yourself from possible fraud on the market, which financial solution would you prefer:

a. Monthly deposit savings into a banks account (interest rates close to zero).
b. Keeping savings at home through the classic "money under the mattress" (with all the risks associated with keeping cash at home).
c. Investing savings in an ethical fund (investment with a high social content and transparency).
d. Allocate savings in an investment fund chosen by you (with the possibility of obtaining returns if the choice proves correct, losses if the choice proves wrong).

(column 2 of Table 5). In particular, the introduction of a *framing effect* determines that the gender and the status do not impact on the probability to invest in an ethical fund whereas age, instruction, yearly income and financial culture impact on the decision to invest ethically. Specifically, the probability to invest in an ethical fund increases for younger and graduated investors whereas it decreases for investors that have a greater financial culture. Finally, investors with a greater yearly income are more ethical than investors with lower income. In this second scenario, the results confirm only our hypothesis $H_{2(a)}$.

4.1.3 Study 3A

To test for the *framing effect*, we propose a third model in which, as dependent variable, we consider the people that changed idea on the investment after adding positive and negative frame applied to our question. The main idea is that the *framing effect* occurs because people, without any specifications (Study 1A), have to make a cognitive effort to evaluate the expected value of each option. In the second question, the options have positive and negative frames and, thus the choice is less costly. The positive frame ("*Investing savings in an ethical fund*") produces a good feeling and people are most likely to choose this option. The contrary in case of negative frames.

If the investor is rational [28], the presentation of logically equivalent options in semantically different ways, would not change the choice to invest in an ethical fund. However, the definition of a framing in positive or negative terms affects people's likelihood of change idea (*attribute framing*).

Table 4 in Panel A reports the change of choices after adding the positive and negative frames. The first evidence is that the inclusion of some positive (negative) frames determine a change in the choices of the investors. In total, 74.2% of the investors changed idea when the question is modified and an attribute framing is introduced. In particular, Panel B of Table 4 shows an increase on the investors that decide to invest in an ethical fund. The investors that, after the introduction of the positive frame, decided to invest in an ethical fund are 39.80% whereas the other solutions record a negative trend.

As in Study 1A and 2A, Study 3A implements the model in Eq. 1 to understand the determinants of the change of the investor's idea. We use as dependent variable, the dummy variable that takes 1 if the investors have decided to invest in an ethical fund after the specification of a positive frame. As in the previous analysis, we use a Probit model to study the determinants.

Column 3 of Table 5 reports the results for the model that consider the probability to change idea when the question adds the positive and negative frames. Specifically, the probability to invest in an ethical fund after adding positive and negative frames depends on gender, age, status, instruction, financial culture and risk appetite. In particular, males are more likely to change idea when the question specify the reason why to save money and the options add negative and positive frames. Furthermore, aged and employed investors are less likely to change idea. Finally, the probability to invest in an ethical fund after adding a positive frame is negatively related to the

Table 4 Change of investor's decision

	Investor's preferences (%)
Panel A	
Investors that do not change idea	25.8
Investor that changes idea (attribute framing)	74.2
Panel B	
Investing savings in an ethical fund (investment with a high social content and transparency)	+ 39.80
Keeping savings at home through the classic "money under the mattress" (with all the risks associated with keeping cash at home)	– 2.6
Allocate savings in an investment fund chosen by you (with the possibility of obtaining returns if the choice proves correct, losses if the choice proves wrong)	– 25.00
Monthly deposit savings into a banks account (interest rates close to zero)	– 12.20

Table 5 Probit model results

	Question 1	Question 2	Change of mind
	(1)	(2)	(3)
Gender	– 0.100*** (0.003)	0.001 (0.006)	0.062*** (0.008)
Age	– 0.132*** (0.037)	– 0.141*** (0.023)	– 0.124*** (0.026)
Status	0.148*** (0.051)	– 0.020 (0.084)	– 0.096** (0.042)
Education	0.506*** (0.127)	0.141*** (0.004)	– 0.158*** (0.006)
Income	0.446*** (0.101)	0.521*** (0.040)	0.319 (0.356)
Financial culture	– 0.062*** (0.004)	– 0.318*** (0.001)	– 0.238*** (0.012)
Tolerance	– 0.018 (0.014)	0.011 (0.019)	0.045*** (0.010)
Intercept	– 1.559* (0.916)	– 0.080 (1.231)	– 0.124 (1.242)
Observations	500	500	500
Pseudo-R^2	0.085	0.059	0.023

Note Estimates report the results of the Probit model
***, **, and * indicate statistical significance at the 1 and 5% levels, respectively

instruction level and the financial culture. Therefore, investors with a higher financial culture are less likely to invest in an ethical fund even if an attribute framing is added to the option. The decision-making process of investors with a high financial culture does not present a framing bias effect. Our findings imply that not all categories of investors are affected by the attribute framing.

4.2 Study B

The purpose of this study is to test the Prospect theory on the ethical investment decision making process. To this purpose, we use three different studies. In a first study, we evaluate the concept of "aversion to loss". In a second study, we use the concept of "aversion to certain loss" by testing the reaction of the investors in relation to the proposal to shift from an investment in an equity fund to an investment in an ethical fund. In these two studies, we focus on the certainty effect. To this purpose we consider how investors perceive risk aversion in choices involving sure losses. Finally, the third study focuses on the isolation effect and evaluate the change of mind when the preferences to the same choice is presented in different forms.

In these scenarios, as in Study A, we test the determinants of the ethical investment decision making process. We specify our model as model in Eq. 1. We apply a Probit model because in all three sub-studies the dependent variable is a dichotomous variable.

4.2.1 Study 1B

We propose a first study in which, as dependent variable, we consider the decision of an investor to maintain the investment after a loss occurs (10% of loss).

Table 6 reports the percentage of investors that decide to remain in the investment and the percentage of investors that, after a loss of 10%, decide to leave the investment. In our sample 45.4% of investors decides to leave the investment whereas 54.6% of the investors decides to remain in the investment.

The model in Eq. 1 considers, as dependent variable, the binary variable taking value one if the investors have decided to remain in the investment, hoping to recover the same loss in a future rise of the financial markets and zero otherwise.

Column 1 of Table 9 reports the results of the Probit model. The probability to remain in the investment, hoping to recover the loss from the raise of the financial market is affected by: gender, age, education and financial culture. Specifically, the Probit model shows a significant and positive sign for the variable age and financial education. This implies that the aged investors and investors with a financial education are less risk averse. Not surprisingly, investors with a financial education have, at least, the basic knowledge to evaluate the future trend of the financial market. In a scenario of loss, they prefer to recover the loss from the future rise of the financial market rather than exit from the investment. Additionally, female e less educated

Table 6 Choices of the investors in Question 3

Options	*Question 3*: You invested 10,000 € in a traditional equity fund three years ago, now due to collapses in the financial markets, you are losing 1000 €, or 10% of your capital; what will you do (%)
Choosing to immediately exit the investment, realizing that it was a bad investment	45.4
Remaining in the investment, hoping to recover the same loss in a future rise of the financial markets"	54.6

investors are more risk averse and thus prefer to exit from the investment in a traditional equity fund. They prefer the certainty of the loss rather than the uncertainty deriving from the future behavior of the financial markets. These results confirm the certainty effect described by Kahneman and Tversky (1979) and support our hypothesis H_5.

4.2.2 Study 2B

The aim of this study is to test the concept of "*aversion to certain loss*". To this purpose, in our question, we describe the situation of an investor that invested 10,000 € in a traditional equity fund and his own position has lost value. In this situation, we propose two options: (a) shift the position to an ethical fund or (b) leave the investment. The aim is to investigate if the decision to leave or to remain in the investment may depends on the specification of the question and options.

Table 7 reports the percentage of investors that decide to shift to an ethical fund and the percentage of investors that, after a loss, decide to leave the investment taking the present value. The summary statistics show that the investors prefer to change the investment and thus, shift to an ethical fund (72% of investors) rather than leave the investment (28%).

To investigate the factors impacting on the decision to shift to an ethical fund, we implement the model in Eq. 1 by using, as dependent variable, the dummy variable taking value one if the investors have decided to shift to the ethical fund and zero otherwise.

The results of Study 2B is reported in Column 2 of Table 9. In a scenario of sure losses, the investment in an ethical fund (without additional costs) appears to be the way to recovery the loss for female, younger and graduated investors. Furthermore, an increase in the yearly income would increase the probability to shift to an ethical fund. The coefficient on financial culture is shown positive and significant. This would imply that the investors with a financial culture are more prone to change the investment from a traditional equity fund to an ethical fund in a scenario of sure losses. A good financial knowledge could reassure investors on the financial

Table 7 Choices of the investors in Question 4

Options	*Question 4*: You started investing your savings three years ago in a traditional equity fund and today your capital should amount to 10,000 € but, today, the value of the fund is 9000 €. Your banker offers you the possibility of transferring your funds from the previous traditional equity fund to a new ethical equity fund without any commission costs (%)
I immediately block the periodic payment to my fund, believing that I can lose even more my savings and I leave the investment taking 9000 €	28
I accept the transition to the ethical fund, hoping that its future rise will make me at least recover its value at the lost moment	72

market trend, making them confident that a subsequent rise will allow them, at least, to recover the lost value. Therefore, well financial educated investors are less risk averse in choices involving possible future gains. Finally, the risk tolerance has a positive and significant impact on the decision to shift from a traditional equity fund to an ethical fund. A high-risk tolerance prevents panic takes over in momentarily unfavorable situations, or perceived as such.

The results of Study 2B corroborate our hypothesis $H_{1(a)}$, $H_{2(a)}$ and H_5.

4.2.3 Study 3B

In this last study, we investigate the factors that may impact the change in the preferences of the investors and the subsequent isolation effect. In particular, we evaluate if there is a switch effect and thus, the investors that do not accept the loss may accept to shift the position to an ethical fund. This would imply that there was a change of mind and so the investors that, in a first scenario, decided to leave the investment in the second situation decide to shift to an ethical fund rather than to leave the investment.

Table 8 shows that 187 investors upon 500 changed the preference and accept the banker's proposal to shift the investment from a traditional equity fund to an ethical fund without additional costs rather than to leave the investment (previous choice).

Table 8 Change of investor's decision

	Investor's preferences (%)
Investors that do not change idea	37.4
Investor that changes idea	62.6

Table 9 Results of Study B

	Question 3	Question 4	Change of mind
	(1)	(2)	(3)
Gender	– 0.359*** (0.041)	– 0.021** (0.010)	– 0.194*** (0.006)
Age	0.121*** (0.030)	– 0.150** (0.064)	– 0.206*** (0.019)
Status	– 0.071 (0.385)	0.239 (0.206)	0.244*** (0.029)
Education	– 0.256*** (0.079)	– 0.122* (0.066)	0.057*** (0.021)
Income	– 0.133 (0.088)	0.141*** (0.044)	0.126*** (0.020)
Financial culture	0.097*** (0.030)	0.114*** (0.024)	0.002 (0.002)
Tolerance	0.091 (0.106)	0.137*** (0.012)	0.009 (0.009)
Intercept	– 0.085 (0.089)	0.208 (0.199)	– 0.444 (1.182)
Observations	500	500	500
Pseudo-R^2	0.0296	0.0213	0.0540

Note: Estimates report the results of the Probit model
***, **, and * indicate statistical significance at the 1 and 5% levels, respectively

To test the determinants of the switch effect, we use the model in Eq. 1. As dependent variable, we use a dummy variable taking value one if the investor changes the preference and zero otherwise.

The change of mind in investment decision making process is affected by gender, age, status, education and income (Column 3 of Table 9).

In Study 1B and 2B, the possibility to recovery the sure loss is uncertain. In the first Study, we indicate the future rise of financial market. In the second Study, we point out the switch to the ethical fund without additional costs. Our findings confirm the isolation effect for some types of investors. Specifically, the investors that are affected by the isolation effect are: female, younger, employed and graduated. For these types of investors, the presentation of the same uncertain scenario in different forms leads to a change of mind in the investment decision making process. However, in Study A, we have demonstrated that the ethical investors have the same characteristics of those that change mind in Study B. So, after a loss occurs, these investors may decide to shift to an ethical fund for moral principles ("*ethics effect*") in addition to the isolation effect.

5 Conclusion

The purpose of this paper is to verify the possible link between the ethical finance choices and the behavioral biases that influence the investor's decision-making process.

Using the survey data of 500 individual investors, we have a number of conclusions. First, the decision to invest in an ethical fund depends on a series of individual factors such as: gender, age, education and employment of investors. Second, the investment decision making process is affected by a framing effect. Therefore, the positive frame added to the option "invest in an ethical fund" lead to a change of mind for 74.2% of the participants. This study adds to the existing literature by showing the investor's characteristics and the behavior biases play in ethical investment decision making process. Our findings show that the decision-making process is affected not only by the option presented in the questionnaire but also by the way that the questionnaire presented the content of the options. Therefore, this study confirms the investors' framing bias in the decision-making process. Third, our study also relates to "prospect theory", as underlined by the aversion to loss bias. Some state of mind such as aversion, loss aversion and mental accounting can affect an individual's decision-making process. After a loss occurs, the investors become less risk averse. Finally, we find evidence of an isolation and ethics effect. This imply that the investors make inconsistent preferences when the same choice is presented in different forms. Nevertheless, the isolation effect is increased by what we call the ethics effects. Investors would change the preferences not only because they are presented in two different forms but also because the investors try to follow their moral principles in making financial investment.

This study also offers practical insights. Considering the increasing number of investors in ethics activities, it is important for financial professionals and policymakers to have a better understanding of the behavioral biases in the ethical investment decision making process The study offers a portrait of the ethics investor in terms of characteristics and behavior on ethical funds. The factors that influence their investment decision making are crucial as these influence their future financial plans. Our findings are useful for financial advisors and institutions to find out the reasons of irrational decision making due to behavioral biases. It will help the researchers to recognize how behavioral biases vary across individualist and moral cultures. It will help the practitioners to understand how the presentation of the investments impact on the decision-making process.

References

1. Agnew JR, Anderson LR, Gerlach JR, Szykman LR (2008) The annuity puzzle and negative framing. Cent Retire Res Brief 8(10)

2. Barnett JH, Karson MJ (1989) Managers, values, and executive decisions: an exploration of the role of gender, career stage, organizational level, function, and the importance of ethics, relationships and results in managerial decision-making. J Bus Ethics 8(10):747–771
3. Browning J, Zabriskie NB (1983) How ethical are industrial buyers? Ind Mark Manage 12(4):219–224
4. Chen C, Tuliao KV, Cullen JB, Chang YY (2016) Does gender influence managers' ethics ? A cross-cultural analysis. Bus Ethics: Eur Rev 25(4):345–362. https://doi.org/10.1111/beer.12122
5. Conroy SJ, Emerson TLN (2004) Ethics and religion : as a predictor of religiosity business ethical awareness students. J Bus Ethics 50(4):383–396
6. Craft JL (2013) A review of the empirical ethical decision-making literature : 2004–2011. J Bus Ethics 117(2):221–259. https://doi.org/10.1007/s10551-012-1518-9
7. Dawson LM (1997) Ethical differences between men and women in the sales profession. J Bus Ethics 16(11):1143–1152
8. Ebrahimi BP, Petrick JA, Young SA (2005) Managerial role motivation and role-related ethical orientation in Hong Kong. J Bus Ethics 60(1):29–45. https://doi.org/10.1007/s10551-005-4962-y
9. Elango B, Paul K, Kundu SK, Paudel SK (2010) Organizational ethics , individual ethics , and ethical intentions in international decision-making. J Bus Ethics 97(4):543–561. https://doi.org/10.1007/s10551-010-0524-z
10. Eweje G, Brunton M (2010) Ethical perceptions of business students in a New Zealand University : do gender , age and work experience matter ? Bus Ethics: Eur Rev 19(1):95–111. https://doi.org/10.1111/j.1467-8608.2009.01581.x
11. Ford RC, Richardson WD (1994) Ethical decision making : a review of the empirical literature. J Bus Ethics 13:205–221
12. Gonzalez C, Dana J, Koshino H, Just M (2005) The framing effect and risky decisions : examining cognitive functions with FMRI. J Econ Psychol 26:1–20. https://doi.org/10.1016/j.joep.2004.08.004
13. Grable J, Lytton RH (1999) Financial risk tolerance revisited : the development of a risk assessment instrument. Financ Serv Rev 8:163–181
14. Grable JE, Lytton RH (2003) The development of a risk assessment instrument: a follow-up study. Financ Serv Rev 12(3):257–274
15. Jones TM (1991) Ethical decision making by individuals in organizations : an issue-contingent model. Acad Manag Rev 16(2):366–395
16. Kahneman D, Tversky A (1979) Prospect theory : an analysis of decision under risk. Econometrica 47(2):263–292
17. Lehnert K, Park Y, Singh N (2015) Research note and review of the empirical ethical decision-making literature : boundary conditions and extensions. J Bus Ethics 129(1): 195–219. https://doi.org/10.1007/s10551-014-2147-2
18. Lei S, Zhang Y (2019) The role of the media in socially responsible investing. Int J Bank Mark. https://doi.org/10.1108/IJBM-09-2019-0332
19. Lewis A, Mackenzie C (2000) Morals, money, ethical investing and economic psychology. Hum Relat 53(2):179–191
20. Linciano N, and Soccorso P (2012) Assessing investors' risk tolerance through a questionnaire. SSRN Electron J
21. Loe F, Mansfield P (2000) A review of empirical studies assessing ethical decision making in business. J Bus Ethics 25:185–204
22. Marino D, Cananzi D, Aragona F (2024) Artificial intelligence and ethics challenges of technologies. In: Ethics and artificial intelligence. Studies in Systems, Decision and Control, vol 519. Springer, Cham. https://doi.org/10.1007/978-3-031-50902-5_3
23. McCabe AC, Ingram R, and Dato-on MC (2006) The business of ethics and gender. J Bus Ethics 64:101–116. https://doi.org/10.1007/s10551-005-3327-x
24. O' Fallon MJ, Butterfield KD (2005) A review of the empirical ethical decision-making literature : 1996–2003. J Bus Ethics 59:375–413. https://doi.org/10.1007/s10551-005-2929-7

25. Rest JR (1986) Moral development: advances in research and theory. Praeger, New York, NY
26. Robin D, Babin L (1997) Making sense of the research on gender and ethics in business: a critical analysis and extension. Bus Ethics Q 7(4):61–90
27. Ruegger D, King EW (1992) A study of the effect of age and gender upon student business ethics. J Bus Ethics 11:179–186
28. Simon HA (1955) A behavioral model of rational choice. Q J Econ 69(1):99–118
29. Thaler RH (2008) Mental accounting and consumer choice : anatomy of a failure. Mark Sci 27(1):12–14. https://doi.org/10.1287/mkscl070.0348
30. Tversky A, Kahneman D (1981) The framing of decisions and the psychology of choice. Science 211(4481): 453–458

The Intricate Balance Between Artificial Intelligence, Ethics, and Algorithmic Standards

Giulio Morabito

Abstract The growing interest in Artificial Intelligence (AI) is linked to the expectation of perfection and responsiveness of these systems, but its diffusion is also associated with imperfections rooted in the nature of systems that reflect human biases. AI, based on *Machine Learning* and *Deep Learning,* can show ambiguities attributable to training data and the 'dark side' of certain behaviours, with different implications. Although these systems are applied in many contexts, some of them present critical issues that need to be seriously addressed due to the ethical, social and legal consequences they may produce. The intent of this investigation is to examine some known cases through their identification from web sources, in order to understand the causes and possible mitigation strategies among those proposed for appropriate use by reducing negative consequences. Every day the press covers significant cases, spotlights shining on major companies that have experimented, perhaps hastily, with applications with resounding failures. Episodes such as the failure of *Microsoft*'s *Tay* chatbot indicate the challenges in dealing with *bias* and discrimination. Even *Syri*, a Dutch anti-fraud system, was suspended for discrimination, highlighting the need to balance privacy and public interest in the digitisation of *welfare*. The case of discrimination by *Amazon*, and *Joule*'s ethical approach, underlines that eliminating *bias* requires continuous efforts. In the field of justice, the assessment of criminal causality entrusted to AI requires impartiality and faces challenges of design and privacy. Awareness of these issues is essential for an ethically responsible development of smart technologies (In this sense, the principles of the European Commission emphasise *fairness*, *transparency* and *explainability* in the use of AI, promoting a responsible and informed approach). However, only the continuous commitment of all those involved can guarantee safety and reliability. While science improves technologies, individual ethics must guide their evolution.

G. Morabito (✉)
DIGIES, Università Mediterranea di Reggio Calabria, Viale dell'Università, SNC, 89124 Reggio Calabria, Italy
e-mail: giuliomorabito97@libero.it

D. Marino and M. Monaca (eds.), *Generative Artificial Intelligence and Fifth Industrial Revolution*, Lecture Notes in Networks and Systems 880,
https://doi.org/10.1007/978-3-031-73880-7_4

Keywords Artificial intelligence ethics · Bias in machine learning · Ethical AI development · AI failures and mitigation · AI in justice systems · Responsible technology deployment

1 Introduction

The advancement of AI technologies has ushered in a new era in automated decision-making processes, from *recruitment* to the complex legal system. The increasing use of these algorithms raises growing concerns about *cognitive biases* affecting the systems, which can significantly distort decisions and contribute to inequalities. In this context, a wide range of studies and research has analysed the impact of *bias* and *opacity* arising from its complexity, exploring how these phenomena affect social, legal and employment dynamics.

The sources reviewed address the issue in the specific contexts of *recruitment* and justice, revealing how such biases can undermine fairness and transparency. Studies such as those conducted by *Gaito*[1] and the considerations of *Psyjob*[2] highlight the need to understand how *biases* can emerge in crucial decision-making areas, underlining how their recognition is crucial to ensure a fairer process.

In recruiting, fuelled by the innovation of AI, studies such as those presented by *I-COM*[3] and *Digital World Italia*[4] illustrate the memorable mistakes and discriminations that can arise during recruitment processes.

In the forensic field, *Gulotta*[5] delves into the impact of *cognitive biases* in the judiciary, while *Durango*[6] offers an in-depth analysis of *racial bias*, highlighting the complexity of the challenges faced in the forensic field.

The investigation extends with articles such as those in *Repubblica*[7] and *Diritto.it*[8] that highlight the adoption of decision-making algorithms in the legal context. Indeed, the case of the wrongful arrest of a person based on an AI error[9] raises crucial questions about the transparency and accountability of such systems. Mitja Gialuz in "Contemporary Criminal Law" addresses the issue of Risk Assessment Tools

---

[1] Www.raffaelegaito.com

[2] Www.psyjob.it

[3] "Bias in artificial intelligence, an open and (still) little discussed issue - I-Com, Institute for Competitiveness" (www.i-com.it)

[4] "Algorithms and AI: 5 memorable mistakes to think about" (www.digitalworlditalia.it).

[5] "General and personal cognitive biases and their impact in the forensic field: foundations and remedies" (www.sistemapenale.it)

[6] 'The Bias of Race - Polarisations of Thought, Identity Twists and the Politics of Hate' (https://air.unimi.it).

[7] www.repubblica.it/economia/2023/07/04/news/intelligenza_artificiale_impatti_lavoro_avvocati_professioni_sentenze-406731550/

[8] 'New scenarios on Artificial Intelligence and Civil Justice' (www.diritto.it).

[9] "In the US they arrested a woman for an AI error" (//ainews.co.uk).

widespread in the United States[10] and Clementina Barbaro in "Questione Giustizia" discusses ethical principles shared at European level[11].

The increasing use of AI in justice systems, with a focus on predictive justice, has generated a debate in Europe. The *CEPEJ* (*European Commission for the Efficency of Justice*) responded with the European Ethics Charter, establishing fundamental and methodological principles. Challenges and cautions were raised on the use of AI, including *risk assessment* tools. The need to respect human rights, ensure transparency and enable user control is emphasised.

In this heterogeneous landscape, the need to promote *transparency* and *explainability* in automated decision-making processes emerges, and these studies help guide the debate and provide ethical solutions to mitigate the effects.

2 Definition and Social Impacts of Bias in AI

AI is increasingly present in our daily lives, contributing to development in various fields thanks to its ability to mimic the functions of the human mind through algorithms. In some cases, this enables autonomous learning, as in *Machine Learning* (*ML)* and particularly *Deep Learning* (*DL*).

The wide range of forms and applications of AI plays an increasingly important role in society, influencing social systems, infrastructures and even individual habits. While its use offers significant advantages in terms of precision and efficient decision-making, there are discriminations associated with this technology. While it promises broader and fairer access to goods and services, there is an increasing need to address the *biases*[12] arising from its use. Discussion on how to mitigate these risks becomes essential to ensure ethical and responsible use in our evolving society.

The term '*bias'* (from the French '*biais*') in the human domain denotes prejudices or erroneous inclinations of the mind. In the decision-making process, the brain employs heuristics as mental shortcuts[13] to facilitate timely decisions, but some of them can lead to errors, originating from mnemonic selectivity, the impulse for quick actions, information overload or lack of data. The concept of *cognitive bias* in humans is closely linked to the survival instinct, but sometimes it manifests distortions in the ability to make objective decisions.

Neural Networks, inspired by the human brain since the 1940s, underpin the mechanisms of *ML* and *DL* and differ significantly from natural neurons. While both act as basic units, the synaptic plasticity of the human brain exceeds the limited

[10] https://archiviodpc.dirittopenaleuomo.org/upload/6903-gialuz2019b.pdf.

[11] https://www.questionegiustizia.it/rivista/articolo/uso-dell-intelligenza-artificiale-nei-sistemi-giud_591.php.

[12] Bias in artificial intelligence, an open and (still) little discussed issue - I-Com, Institute for Competitiveness

[13] Raffaele Gaito 'Cognitive biases: what they are and how to recognise them' (www.raffaelegaito.com).

adaptability of artificial neural networks. Artificial neurons are interconnected in layers, but the human brain shows excellent energy efficiency and handles complex tasks, in contrast to artificial neural networks that require considerable computational resources, highlighting the challenges in replicating intuition, emotion and decisions based on various stimuli. These networks analyse data, learning and recognising patterns. *DL* is a subcategory that uses *deep neural networks* with multiple hidden layers for complex tasks such as image recognition and natural language. The algorithm evaluates the accuracy of the results using its own network of neurons without human involvement in the review. During training, the network progressively learns from both the input data and its own *feedback*, autonomously demonstrating its ability to learn over time.

In *ML* models, the learning required for training[14] the network involves different approaches depending on human involvement and the type of data, resulting in four different categories: supervised, semi-supervised, unsupervised, or by reinforcement.

In general, *ML* uses data to identify patterns and make predictions, while *DL* extends these capabilities without requiring direct human review, through training and inference steps, making accurate predictions on new data.

Algorithms, therefore, learn from training data, and this may be reflected in the biases present in society. For example, if the historical data used to train a personnel selection system are based on discriminatory decisions or reflect inequalities in the past, the algorithm may perpetuate these inequalities. Many databases have little heterogeneous characteristics that lead to biased learning. For example, 99% accuracy in facial recognition algorithms has been observed for white men and 35% for dark-skinned women.

In criminal proceedings, the use of AI is growing rapidly, with applications both in *predictive policing* and in evidentiary settings, with the increasing use of algorithmic evidence.

In America, several states have regulated the use of risk assessment in the pre-trial process, with *risk assessment* mechanisms also used in the sentencing phase, with widespread use in the juvenile justice system. However, several errors have been identified, such as a notorious case that led to the arrest of an African-American woman in her eighth month of pregnancy, due to misidentification by a system in comparing images of the real suspect with a photograph in the police database.

In the field of justice, the need for certainty, speed and bureaucratic simplification in investigative and judicial activities provides a push towards the use of intelligent systems, whose potential *biases* conflict with the human interpretation required in law. The use of this technology as a support, reflects the dynamic nature of legal issues, but there are challenges related to transparency, risks of discrimination and limitations in analysing human elements affected by *cognitive bias*.

In determining criminal causality, alternative theories in forensic practice mitigate the limits of causal equivalence, but defining the degree of certainty is problematic. The use of AI in the justice system aims at 'computable', but impartiality, careful

[14] Step in which input data are associated with weights.

evaluation of potential design flaws, as well as misuse of data and threats to *privacy* must be ensured.

AI is also applied in the personnel search and selection process, a crucial decision-making phase susceptible to various *biases*, related to personal characteristics (*primary bias*) as well as related to behaviour and socio-economic variables (*secondary bias*).

Some very common ones include the *contrast effect*[15], *the primacy effect*[16] or the *recency* effect[17] and the *halo* effect[18], which can lead to biased evaluations of candidates.

These *biases* can jeopardise the success of *recruitment* actions.

Kriti Sharma, an Indian ML expert, reveals that when she participates in online forums with her name and image, she is often asked for references, while in the absence of gender indications this never happens. These cases highlight the need to address prejudices and carefully consider the ethical and social implications of such technologies. Gender discrimination affects remuneration, job opportunities and job advertisements, contributing to entrenched prejudices in society[19].

A well-known case is related to an *ML-based recruiting* engine developed in 2014 by *Amazon* researchers. The aim was to automate its processes for finding the best talent by training a model with data from resumes sent to Amazon over a 10-year period.

Despite the failures, the desire of employers remains to improve this process and more and more companies are investing in intelligent systems.

Recently, a recent economics graduate founded and chairs a start-up called '*Joule*', conceiving a platform with the intention of revolutionising the labour market landscape. This innovation is based on principles of synergistic integration between human intrinsic value and the power of AI, "*aiming to overcome the limitations of traditional models*". The platform aims to optimise human resources identification processes and to facilitate the match between companies and aspiring employees through the assessment of personal skills and individual talents (*Soft Skill*[20]). The implemented application procedure integrates a video presentation and other information then processed by the IA.

---

15 " ... *the answer given to a question is in the opposite direction to the answer given to a previous question to a greater extent than would occur.*" (www.tesionline.it/appunti/psicologia/tecniche-dell-intervista-e-del-questionario/contrasto-e-assimilazione-e-i-processi-di-inclusione-ed-esclusione-delle-informazioni-attive/501/40)

16 '*The information we acquire first guides the interpretation of subsequent information*' (www.skuola.net/psicologia/effetto-primacy.html)

17 "... identifies the cognitive bias whereby it is easier to remember the last items on a list of information that needs to be memorised." (https://blog.blumatica.it/effetto-primacy-effetto-recency-bias-sicurezza-sul-lavoro/)

18 "... is a biascognitive whereby the perception of a trait is influenced by the perception of one or more other traits of the individual or object. An example is judging a good-looking individual to be intelligent at first glance." (en.wikipedia.org/wiki/Effect_alone)

19 www.amministrazioneincammino.luiss.it/wp-content/uploads/2020/06/ZUDDAS.pdf

20 https://www.almalaurea.it/info/aiuto/lau/manuale/soft-skill.

Joule's own *CEO* states that the algorithms can reveal certain discriminatory predispositions and to mitigate this risk, a team of experts specialised in identifying and correcting the *biases* that emerge has been set up to refine the learning models.

3 Bias and Opacity

ML takes the form of a set of models that allow a machine to learn from data without the need for explicit programming for a specific task. This approach departs from the original symbolic model, characterised by the prior definition of rules by the programmer, in that it provides a method of learning applied to data. This process evolves so that the system can improve its performance on a specific task, gradually gaining experience from the data and allowing the machine to autonomously extract specific decision rules for particular cases [5].

Recent developments point to advanced capabilities in generative AI, ranging from fluid conversation to generating news articles or writing code. Current success is intrinsically linked to the speed of computation, the use of *neural networks* and the wide availability of large data sets (*big data)* enabling complex tasks to be tackled.

Despite this, fundamental questions arise about the very definition of intelligence and the ways in which AI achieves its goals, aspects that are still unclear and can be defined as system *opacity* or *black box.*

This opacity gives rise to a complex ethical-legal dilemma associated with such systems, in which decision rules are autonomously extracted through the analysis of *big data.*

This is recognised by the scientist *Yann LeCun* himself, a key figure in the development of *convolutional neural networks*[21]. The lack of transparency is attributable to the complexity of its operations and parameters, which are beyond human comprehension. Although this complexity offers advantages in terms of operational efficiency, it raises significant ethical and legal issues. AI decisions may be difficult to understand or accept from a legal point of view.

The complete comprehensibility of an algorithm is not always guaranteed and may be hindered by various circumstances such as the general complexity of the algorithms, the logic of which is often difficult to explain, giving rise to a 'technical opacity' that may elude those without specific expertise. It should be noted that even if the source code were to be fully disclosed, the understanding of the complete functioning of the machine could remain obscure, as the machine learning process is often unknown to the developers themselves.[22]

[21] "A convolutional neural network can have dozens or hundreds of layers, each of which learns different features of an image. Filters at different resolutions are applied to each training image and the output of each convolutional image is used as input for the next layer. The filters can start with very simple features, such as brightness or edges, and become increasingly complex until they include features that uniquely define the object." (Source: https://it.mathworks.com/discovery/convolutional-neural-network-matlab.html)

[22] https://www.amministrazioneincammino.luiss.it/wp-content/uploads/2020/06/ZUDDAS.pdf.

The relevance of the opacity issue is not limited to the complexity of operations and parameters beyond human comprehension, but extends its scope to questioning the very meaning of the concept of knowledge. Although opacity enables AI to carry out complex tasks quickly, as in the case of designing robots or determining the structure of proteins, some authors are optimistic that it can reduce or eliminate the need for theories. This perspective presents itself as a potential model for a new scientific paradigm that does not require in-depth understanding.

In fact, the principle of '*algorithmic transparency*' requires entities that take decisions by means of automated data processing systems to provide comprehensible explanations of the procedures and to justify the decisions. This responsibility is particularly relevant for public entities and in cases where computer systems are involved in judicial and administrative decisions.

In the judicial context, the lack of transparency would violate the duty to *state reasons in court* and the right to defence.

Administrative decisions also require comprehensible explanations of the criteria used by the algorithm employed. *Algorithmic transparency* is considered crucial for preserving participation, access and motivation in public decisions, as highlighted in a number of legal rulings concerning the extraordinary mobility procedures of public school teachers. The jurisprudence emphasises that the absence of motivation undermines the right of action and defence in court, making it necessary for the administrative judge to have full cognisance and review of the '*algorithmic rule*'.

No system can be trained to deal with every possible new case based only on previously observed ones, as the long tail theory might suggest[23]. Although AI can undergo additional training cycles to reduce errors, human supervision still determines what is considered an error and what is not.

4 Solutions and Approaches to Reduce Bias and Opacity

The black box definition of machine learning systems is more a metaphor than a technical reality. Such systems are composed of multiple elementary reactions, but the complexity of the interactions makes it difficult for humans to understand the decision-making process.

There are two main ways of interpreting explanation in this context: one focusing on the reason for an early decision and the other looking for a retrospective explanation in the case of failure. In AI, an explanation of the first type is often expected, but ML systems often operate more like the second type, reacting to stimuli without explicit reasoning. Difficulties in understanding and reconstructing decisions are considered practical challenges rather than insurmountable mysteries.

[23] https://www.problemsetting.it/atlante-di-problem-solving/marketing/coda-lunga/#:~:text=The%20theory%20of%20the%20long%20tail%20for%20a%20company%20in%20a%20region.

In the context of technology, the issue of bias remains unresolved, but there are several solutions to mitigate its harmful consequences[24]. One initial strategy is to anonymise documents by removing names and gender pronouns, and to exclude explicit references to minorities or protected classes. For DL-based applications, the technology itself can be used to detect bias by means of *Adversarial Learning*[25]. In addition, an effective practice is 'data *cleaning'*, i.e. the filtering of training data, an operation based on reducing human discriminatory behaviour to achieve less biased patterns.

In the field of *recruiting*, a promising approach to mitigate cognitive biases could be '*blind recruiting*'[26].

The '*debiasing*' process should not be seen as an isolated act, but rather as an ongoing commitment to excellence. Experts and programmers, together with consumers, must actively detect signs of bias, implement review processes and consider special cases.

Currently, '*hybrid*' systems that combine human judgement with AI are reported to be highly effective in reducing bias and ensuring better results. While algorithms bring rigour and repeatability, human input offers a more contextualised assessment, adding awareness, research expertise and understanding of decision-making[27].

In the legal and administrative context, it is important to distinguish between the process of exploring and searching for a solution and the moment of justifying a decision. Control over the decision must remain in the hands of human beings who must justify their choices. Responsibility for the decision and justification must remain human[28].

Padua[29], suggests precautions such as strengthening the judge's free conviction and reinforcing the duty to state reasons in order to preserve justice and fairness in the process.[30]

The *principle of fairness* aims to ensure that the data used in training is inclusive and predictable, thus promoting impartial treatment. Transparency is also fundamental, requiring AI systems to record and document decisions and the logical path followed to reach them. Furthermore, the principle of *explainability* aims to provide comprehensible information to users about the impact and decision-making process and is closely related to that of *interpretability*, although the latter refers to the model's ability to determine cause and effect.

Algorithmic transparency can be examined through four key aspects:

---

[24] Bias in artificial intelligence, an open and (still) little discussed issue - I-Com, Institute for Competitiveness

[25] en.wikipedia.org/wiki/Adversarial_machine_learning.

[26] www.in-recruiting.com/en/blind-recruitment-how-anonymous-cvs-improve-recruitment-and-reduce-cognitive-bias/

[27] www.i-com.it/2022/01/14/bias-intelligenza-artificiale/

[28] www.agendadigitale.eu/cultura-digitale/bias-rumore-E-I-missing-friends-ecco-perche-alcune-tecnologie-sono-piu-neglette-di-altre/

[29] Giorgia Padua PhD in Public Law—'Tor Vergata' University of Rome.

[30] 'Artificial intelligence and criminal judgement: scenarios, limits and perspectives' (www.processopenaleegiustizia.it).

The presence of an automated decision-making process, its relevance, the functioning of the algorithm and the consequences for the individual. The first aspect emphasises the right to be informed about the presence of an automated decision-making process, including the obligation to inform the individual about the use of an algorithm to process his or her personal data. The second aspect concerns transparency about the significance of the algorithm in automated decision-making and its consequences for the individual, as set out in Article 22 of the GDPR.

The principle of transparency extends to the individual's right to obtain meaningful information about the logic used in the automated processing of his or her data. This goes beyond simply being aware of the existence of automated processing, requiring an understanding of the logic of the algorithm. The GDPR specifies that this information must be '*meaningful*' and make it understandable how the algorithm works.

The principle of '*knowability*' of the algorithm implies a complete understanding of the '*robotized decision*', including authors, processing, decision mechanism and relevant data. This knowledge can be obtained through a proactive approach during programming or a reactive approach involving access to the software source code, although complete understanding of the algorithm may not always be guaranteed due to technical opacity, ML complexity and legal reasons.

However, the right to know how the algorithm works is not always sacrificed and the data controller must provide that information that is not related to the protection of industrial or corporate secrets. Since trade secret protection may conflict with other legally protected interests, a balancing of opposing interests is appropriate even if it entails sacrificing the legal positions of software owners.

Lorenza Saettone highlights the influential role of women such as Joy *Buolamwini*[31] and Shalini Kantayya[32] in highlighting AI-related discrimination. *Buolamwini* conducted a study called '*Gender Shade*' on facial recognition bias, leading companies such as *IBM* and *Microsoft* to block the sale of such systems to law enforcement agencies. *Kantayya's* documentary '*Coded Bias*', inspired by *Buolamwini*, explores ML biases, highlighting how AI reflects human *biases*, including racism and sexism in society's data.

AI, through the example of facial recognition, used in areas such as recruitment and insurance, can lead to misinterpretations of minorities, contributing to insidious forms of segregation.

Saettone believes that AI should be considered a metaphor for ourselves, as it reflects human characteristics and cruelties.

---

[31] Computer scientist at the MIT Media Lab and founder of the *Algorithmic Justice League*.

[32] American filmmaker and environmental activist whose films explore human rights.

5 Case Studies

Tay, a chatbot developed by Microsoft and launched on Twitter in March 2016, was an experiment to test human-artificial interaction and natural language learning capabilities. Designed as a teenage girl, Tay used machine learning to acquire knowledge from online conversations. However, the experiment quickly ended in unforeseen failure when, a few hours after launch, she began posting tweets containing racist, misogynistic and anti-Semitic content. This behaviour was attributed to the chatbot learning from negative interactions of Twitter users, who provided information and suggestions negatively influencing its behaviour.

Faced with the inappropriate content, Microsoft quickly suspended the service to make changes, but controlling Tay's behaviour proved difficult and eventually Microsoft opted to permanently deactivate the service. This incident highlighted the challenges in training algorithms, especially to deal with bias and discrimination, especially when exposed to unmoderated interactions.

Syri, the Dutch anti-fraud system of 2014, raised concerns about its disproportionate use. The *Court in The Hague* found a lack of balance between privacy and public interest. The algorithm charged 26,000 households without specific suspicion, violating European privacy regulations and discriminating by wealth and ethnicity. This case reflects a global trend of integrating AI into welfare, but *Alston*[33] warns that meaningful consultation with beneficiaries is often lacking, threatening privacy and access to benefits. *Syri's* suspension is a turning point in the rights-based resistance against digital surveillance in welfare. Despite the possibility of an appeal, *Alston* trusts that this decision will alert politicians to the challenges of digitising welfare.[34]

COMPAS (*Correctional Offender Management Profiling for Alternative Sanctions*) algorithm used in some US court systems to estimate a defendant's likelihood of reoffending has aroused intense discussion due to its discriminatory tendency.

The main problem with COMPAS emerged from the revelation of a marked bias in the system, which resulted in a significantly higher number of false positives for recidivism in the case of black offenders (at a rate of approximately 45%), than for Caucasian defendants (where the system recorded a rate of 23%). This phenomenon, known as 'racial disparity', has raised significant concerns about the fairness of the justice system[35].

Like many similar cases, the bias present in COMPAS can be attributed to several elements, including the data used to train the algorithm, the choice of *ML* model, and the process of creating the algorithm itself. Accurate analysis of these factors is essential to ensure that the algorithms used in judicial systems are fair and free of unfair discrimination.

---

[33] Philip Alston, UN Special Rapporteur on extreme poverty and *human* rights.

[34] www.osservatoriodiritti.it/2020/02/18/intelligenza-artificiale-applicazioni-pratiche-nella-vita-quotidiana-esempi/

[35] Bias in artificial intelligence, an open and (still) little discussed issue—I-Com, Institute for Competitiveness

Amazon developed an AI-based selection tool. This system, designed to facilitate the staff recruitment process, came under scrutiny because of its alleged discrimination against female candidates. The selection system used AI to analyse CVs and identify the most suitable candidates for certain positions. However, the problem emerged when it was discovered that the system had been trained mainly on data from male candidates. As a result, the algorithm showed a propensity to automatically favour male candidates and discriminatorily exclude female candidates.

Amazon acknowledged the critical issues and limitations of the system, admitting that the algorithm was not neutral and reflected inequalities in the training data. Information suggests that the company abandoned the project in 2018, probably due to concerns about gender discrimination and lack of fairness.

This incident highlights the challenges and responsibilities associated with the development and implementation of algorithms in sensitive contexts such as human resources management. The incident contributed to the broader debate on ethics in the use of AI and led to a greater focus on the importance of training algorithms on diverse and representative data in order to avoid discrimination and ensure impartiality in automated decision-making.

Joule

The process of finding and hiring a non-management employee involves an average cost of three months and a fee of 20% of the Gross Annual Income (RAL), with a failure rate ranging between 15 and 25% in the first three years. In Italy, spas have invested more than EUR 1 billion in recruiting new business figures. However, this market could undergo a revolution with the adoption of AI.

The Padua-based *start-up*, founded by recent graduate *Laura Sposato*, is introducing Joule, an AI algorithm for personnel selection. The system promises to significantly reduce the time and costs associated with finding and hiring new employees. *Joule* analyses hundreds of physical and psychological variables in just forty seconds, providing a response that could revolutionise the traditional personnel selection process.

The *start-up has raised* funding from a group of 9 *business angels*, proving that innovative ideas can also find support in Italy. *Laura Sposato* explains that Joule values candidates based not only on skills, but also on '*soft skills*', i.e. personal qualities that are useful in the work environment. The machine objectively measures these characteristics, reducing the probability of a candidate's 'rejection' in the new work environment. Joule aims to make the selection process fast and effective, with an expected success rate of more than 95% thanks to the heterogeneous team that is committed to correcting *bias on a* daily basis.[36]

[36] www.padovaoggi.it/economia/joule-la-start-up-dei-millennials-trova-il-lavoro-ideale-in-40-secondi.html

6 Future Considerations

In the perspective of shaping an equity-driven future in AI, several approaches and initiatives of crucial relevance emerge. It is imperative that developers adopt ethical practices early in the design process. This requires the implementation of '*debiasing*' policies in training data and compliance with ethical standards throughout the lifecycle of such systems.

Maximum transparency in algorithmic decision-making processes, together with the provision of clear and comprehensible explanations of the impact and reasoning behind decisions, is an essential element in instilling trust and facilitating the accurate evaluation of resolutions.

The establishment of strict regulations and standards is another crucial bulwark. Specific regulations on use in critical sectors, such as health, education and justice, can provide a framework to mitigate the risks associated with bias [2, 4, 10, 11].

Compliance with regulations, including those already in force, remains imperative; it is necessary to ensure that technologies comply with privacy, non-discrimination and civil rights laws in order to avoid unintended legal consequences and protect individuals from unjustified discrimination.

International collaboration, through the adoption of global guidelines and the sharing of best practices worldwide, is an essential requirement, considering the transnational scope of the ethical issues involved [1, 3, 6, 7].

Inclusiveness is also of particular importance and can be achieved through the diversification of those involved in the development, with the participation of heterogeneous teams, acting as a key strategy to avoid inherent bias and address a broader range of perspectives [8, 9].

A further critical aspect is the indispensable human involvement. AI systems must be designed to complement human judgement, allowing a constant exchange with human experts in order to improve understanding of the context and correct any distortions or wrong decisions.

Supporting the development of new methods and approaches to mitigate bias is essential and requires an ongoing commitment of the science and technology community.

Finally, public awareness plays a key role, and there is a need to educate users on the use of these systems, explaining the risks associated with biases and the potential benefits, helping to form a more aware and critical user base.

References

1. Cuomo C, Marino D (2019) The local work plans (LWP) and territorial economic system (TES): assessment and evaluation. Studies in systems, decision and control, vol 180, pp 225–237. ISSN: 2198-4182. https://doi.org/10.1007/978-3-030-00677-8_19
2. Dattola S, Morabito FC, Mammone, N, La Foresta F (2020) Findings about loreta applied to high-density EEG-a review. Electronics 9. ISSN: 2079-9292

3. Foti A, Marino D (2020) Blockchain and charities: a systemic opportunity to create social value. In: Marino D, Monaca M (eds) Economic and policy implications of artificial intelligence. Studies in systems, decision and control, vol 288. Springer, Cham. https://doi.org/10.1007/978-3-030-45340-4_11
4. Mammone N, De Salvo S, Marino S, Bonanno L, Ieracitano C, Dattola S, La Foresta F, Morabito FC (2020) Estimating the asymmetry of brain network organization in stroke patients from high-density EEG signals. Springer, Singapore. ISBN: 978-981-13-8949-8
5. Marino D, Cananzi D, Aragona F (2024) Artificial intelligence and ethics challenges of technologies. In: Ethics and artificial intelligence. Studies in systems, decision and control, vol 519. Springer, Cham. https://doi.org/10.1007/978-3-031-50902-5_3
6. Marino D, Tebala D (2021) Economy and development in Calabria: the weak development hypothesis. Rivista Internazionale di Scienze Sociali–anno CXXIX 2:187–22, ISSN 0035-676X. https://doi.org/10.26350/000518_000068
7. Marino D, Gil Lafuente J, Tebala D (2023) Innovations and development of artificial intelligence in Europe: some empirical evidences. Eur J Manage Bus Econ 32(5):620–636. https://doi.org/10.1108/EJMBE-03-2023-0085
8. Marino D, Tebala D (2022) Rural areas and well-being in EU countries + UK: a taxonomy and a cluster analysis. Sustainability 2022(14):15213. https://doi.org/10.3390/su142215213
9. Tebala D, Marino D, Bianchino A (2023) A provincial indicator to describe potential artificial intelligence. In: Marino D, Monaca M (eds) Artificial intelligence and economics: the key to the future. Lecture notes in networks and systems, vol 523. Springer, Cham. https://doi.org/10.1007/978-3-031-14605-3_1
10. Versaci M, di Barba P, Morabito FC (2020) Curvature-dependent electrostatic field as a principle for modelling membrane-based MEMS devices. A review. Membranes 10
11. Zhang YD, Morabito FC, Shen D, Muhammad K (2021) Advanced deep learning methods for biomedical information analysis. Neural Networks 133:101–102

Sitigraphy

12. www.raffaelegaito.com/bias-cognitivi-esempi/#:~:text=Per%20giustificare%20e%20dare%20un,il%20nome%20di%20bias%20cognitivi
13. www.psyjob.it/balzare-alle-conclusioni-bias-cognitivi-e-complottismo.htm#:~:text=Fortunatamente%2C%20%C3%A8%20possibile%20allenare%20le,pregiudizi%20e%20a%20imparare%20da%20essi
14. www.sistemapenale.it/pdf_contenuti/1623396735_gulotta-2021b-bias-cognitivi-campo-forense.pdf
15. https://air.unimi.it/bitstream/2434/604757/2/BIAS%20DELLA%20RAZZA%20-%20DURANGO%20EDIZIONI.pdf
16. www.digitalworlditalia.it/tecnologie-emergenti/intelligenza-artificiale/algoritmi-e-ai-cinque-memorabili-errori-su-cui-riflettere-131932
17. www.agendadigitale.eu/cultura-digitale/assunzioni-tramite-ia-e-discriminazioni-contro-i-disabili-gli-usa-corrono-ai-ripari/
18. https://inda.ai/it/bias-e-intelligenza-artificiale-qual-e-limpatto-sul-recruiting/
19. https://forbes.it/2023/10/05/joule-startup-intelligenza-artificiale-lavoro/
20. www.repubblica.it/cronaca/2022/08/26/news/addio_curriculum_grazie_alla_mia_app_troverete_lavoro_in_40_secondi-362939247/
21. https://www.peoplechange360.it/people-strategy/development-and-learning/soft-skill-che-cosa-sono-perche-svilupparle/
22. www.altalex.com/documents/news/2020/08/03/intelligenza-artificiale-euristica-e-bias-cognitivi-applicati-agli-algoritmi
23. www.altalex.com/documents/news/2023/09/27/intelligenza-artificiale-processo-importanza-fattore-umano

24. https://www.repubblica.it/economia/2023/07/04/news/intelligenza_artificiale_impatti_l avoro_avvocati_professioni_sentenze-406731550/
25. www.diritto.it/nuovi-scenari-su-intelligenza-artificiale-e-giustizia-civile/
26. //ainews.it/in-usa-hanno-arrestato-una-donna-per-lerrore-di-unai/
27. https://archiviopenale.it/File/DownloadArticolo?codice=bf1948e5-bcb4-46c7-a4cc-9d0001 64350c&idarticolo=39500
28. https://ristretti.org/lintelligenza-artificiale-che-condanna-negli-usa-e-realta-litalia-la-usa-solo-per-il-riconoscimento-facciale
29. https://discrimen.it/wp-content/uploads/Carlizzi-Il-principio-del-libero-convincimento.pdf
30. www.digital4.biz/marketing/intelligenza-artificiale-e-digital-marketing-come-difendersi-dai-bias/
31. www.digital4.biz/marketing/big-data-e-analytics/intelligenza-artificiale-nel-marketing-que llo-che-i-cmo-devono-sapere/
32. learn.microsoft.com/it-it/security/engineering/failure-modes-in-machine-learning
33. www.diariodiunanalista.it/posts/introduzione-alle-reti-neurali-pesi-bias-e-attivazione/#:~: text=Una%20rete%20neurale%20%C3%A8%20in,approccia%20ad%20una%20soluzione% 20accettabile.
34. https://it.mathworks.com/discovery/convolutional-neural-network-matlab.html
35. https://www.ibm.com/it-it/topics/data-labeling
36. www.skuola.net/psicologia/effetto-primacy.html
37. https://blog.blumatica.it/effetto-primacy-effetto-recency-bias-sicurezza-sul-lavoro/
38. it.wikipedia.org/wiki/Effetto_alone
39. https://www.coe.int/en/web/cepej
40. https://archiviodpc.dirittopenaleuomo.org/upload/6903-gialuz2019b.pdf
41. https://www.questionegiustizia.it/rivista/articolo/uso-dell-intelligenza-artificiale-nei-sistemi-giud_591.php
42. https://www.almalaurea.it/info/aiuto/lau/manuale/soft-skill

Smart Citizens and AI: Trusting the AI

Tommaso Marino, Luca Ballestra Caffaratti, Paola Pecchio, Dino Galante, Andrea Piccione, and Francesca Galeno Mellucci

Abstract Technological advances and the widespread adoption of Artificial Intelligence (AI) have sparked a revolution with profound social and personal implications. The proliferation of AI devices, involving millions of users and generating valuable feedback for model optimisation, highlights the continuing expansion of this phenomenon. The impact of AI in education is particularly significant, where its applications promise to transform learning and teaching. However, the current level of understanding and adoption of AI for educational purposes (AIED) is limited, and most teachers lack adequate training to use it effectively. The aim of this article is to explore how the use of AI devices in schools can support student learning, to analyse the reliability and effectiveness of these tools in solving logical–linguistic and scientific problems, and to discuss how teachers can use AIED in a pedagogically aware way.

Keywords Smart citizen · Artificial intelligence · Social implication of AI · Educational use of AI

T. Marino (✉)
Associazione per l'Insegnamento della Fisica AIF, Turin, Italy
e-mail: tommasomarino@gmail.com

L. B. Caffaratti
Università degli Studi di Torino, Turin, Italy

P. Pecchio
I.C. 66 Martiri Grugliasco, Turin, Italy

D. Galante
IIS Pascal, Giaveno (TO) E AIF Torino, Turin, Italy

A. Piccione
EFT-USR Piemonte, Corso Vittorio Emanuele II, 70, Turin, Italy

F. G. Mellucci
IIS Curie-Levi, Collegno (TO), Turin, Italy

D. Marino and M. Monaca (eds.), *Generative Artificial Intelligence and Fifth Industrial Revolution*, Lecture Notes in Networks and Systems 880,
https://doi.org/10.1007/978-3-031-73880-7_5

1 Introduction

Technological advances and the widespread adoption of Artificial Intelligence (AI) have sparked a revolution with profound social and personal implications. The proliferation of AI devices, involving millions of users and generating valuable feedback for model optimisation, highlights the continuing expansion of this phenomenon. The impact of AI in education is particularly significant, where its applications promise to transform learning and teaching. However, the current level of understanding and adoption of AI for educational purposes (AIED) is limited, and most teachers lack adequate training to use it effectively. The aim of this article is to explore how the use of AI devices in schools can support student learning, to analyse the reliability and effectiveness of these tools in solving logical–linguistic and scientific problems, and to discuss how teachers can use AIED in a pedagogically aware way.

In this paper we present some considerations about the use of AI systems in schools, accompanied by a description of an experience carried out in the classroom with students. It is also mentioned here the possibility of implementing a teaching unit using a machine learning which, starting from a dataset of images for galaxy recognition, can accompany students to familiarize themselves with these technologies.

The dataset cited here:

(https://astronn.readthedocs.io/en/latest/galaxy10sdss.html) is one of many available open on the Internet network. This allows platforms to be trained to build AI systems for use in the classroom with students, to accompany the study of the topics covered by the dataset (in this case astronomy).

2 Literature Review

The past decade has seen an exponential growth in public and private investment in research and innovation in the field of artificial intelligence (AI) [20]. One of the most disruptive outcomes of this process was the experimental release of ChatGPT, the first public and free application of a large language model (LLM). This event marked the definitive entry of AI into the public debate. A key element of LLMs is the ability to process natural language (NLP) in written and spoken form, enabling the understanding of human language input and the production of natural language responses [9]. Following the release of ChatGPT, numerous IT companies have begun to release their own LLM applications, including Google Bard, Microsoft Bing and Adobe Firefly. The proliferation of AI systems is leading to an exponential increase in human–machine interactions in various areas of social and individual life [27]. The rapid development of technology and the massive use of AI devices in everyday life are laying the foundations for a technological revolution with significant implications at the social and personal level [1]. We are currently witnessing a proliferation of AI devices involving millions of people. The use of these systems generates valuable feedback that helps to improve the models on which they are based.

AI is a rapidly evolving field of computational science that seeks to develop tools and techniques for solving problems that are believed to require human intelligence [16]. Although artificial general intelligence (AGI) comparable to human intelligence has not yet been achieved [18], so-called 'weak AI' is already having a significant impact on individuals, organisations and societies [4]. Weak AI focuses on systems that replicate human mechanisms and behaviour in specific tasks, "such as visual perception, speech recognition, decision-making, and problem-solving, by using algorithms, data, and computational power to simulate human-like intelligence in machines" [10]. One of the key elements that characterize AI is Machine Learning (ML), a statistical approach that allows machines to be progressively trained to solve specific problems based on the data available and the experience accumulated in solving them [15].

The still ambivalent transformative potential in different sectors of society led the European Council in December 2023 to formulate a proposal that aims, on the one hand, to stimulate investment and innovation in AI and, on the other hand, "promotes so-called regulatory sandboxes and real-world-testing, established by national authorities to develop and train innovative AI before placement on the market" [6]. Recently, there is a growing interest in the use of artificial intelligence in education (AIED). Ullrich et al. [22] highlight that the publication of researches and articles on AIED has gradually increased since 2017, particularly in the United States and China. The authors argue that the Covid-19 pandemic has accelerated the process of digitisation of education, driving the adoption of technologies in education (EdTech), and that ultimately the emergence of AI systems capable of human-like performance in various areas such as learning, adaptation, and data synthesis has created new challenges and opportunities for the integration of AI in EdTech [22]. The focus on AIED has also involved major international organisations. Between 2019 and 2021, UNESCO published three documents [23–25] highlighting the enormous potential of AI to achieve Goal 4 of Agenda 2030 (Quality Education). However, the same documents also highlight the risk that AI may herald new educational inequalities [7].

Despite the growing interest in AIED, knowledge in this area is still limited and most teachers lack the necessary mastery to use AI effectively in the classroom [8]. Furthermore, there is a lack of specific training on AIED, especially in the pre-service phase [21]. However, it is precisely this phase that is crucial for the development of positive attitudes towards AIED among teachers, a key factor for effective implementation in the classroom. Some scientific studies have analysed the potential impact of AIED on education. Xia et al. [28] identify 13 roles of AI in 4 key areas: "learning, teaching, assessment and administration" [28]. AI can influence student learning in terms of "motivation, engagement, academic performance, twenty-first century skills and non-cognitive aspects" [28]. It can also support teacher practice and training in terms of "work efficiency, teaching skills and attitudes towards AIED" [28]. However, the introduction of AI in EdTech poses several challenges for the future. Zawacki-Richter et al. [30] highlight the lack of theoretical analysis from a psycho-pedagogical perspective in the field. The prevailing approach is an engineering one, focusing on the development of predictive models through empirical research.

From a psycho-pedagogical perspective, a particularly relevant issue is the level of "agency" [2] that should characterize the interaction between students, teachers, and AI. Brod et al. [3] propose a model with four levels of autonomy and control of the learning process: full student control; shared control between student and AI; full teacher control; shared control between teacher and AI. The authors highlight also the potential of AI to develop an adaptive assignment of the level of agency based on user characteristics.

Beyond the use of AI systems for specific tasks, a growing line of research focuses on the characteristics of the relationship between humans and emotionally intelligent artificial intelligence (EAI) systems [27]. According to Weber-Guskar, these systems, such as animal robots, chatbots and holograms, are designed to recognise and simulate typical human emotional responses, creating more engaging and personalised interactions in education. Experimental research also shows that the use of virtual assistants can help learners compensate for deficits in self-regulation skills by providing adaptive support to learners' skills by modulating their behaviour according to individual needs [13]. According to Sun et al. [5], self-regulation strategies can also be enhanced through the use of learning platforms integrated with AI devices. These devices provide real-time feedback and allow for graphical visualisation of progress, which motivates continuous learning and allows learners to instantly visualise learning outcomes through radar charts.

Although still a developing field, some studies have highlighted the potential benefits of AI in improving the education of students with disabilities and special educational needs (SEN). Currently, AI is a little-used technology for the development of supports specifically settled for people with disabilities. The scientific literature on the subject is limited, and there is a lack of specialised training programmes for practitioners and regulatory measures to protect the most vulnerable users. However, some authors envision the potential benefits of AI in this area: Smith et al. [19] hypothesize the use of AI for the development of customised assistive technologies for people with disabilities. Lampos et al. [11] suggest how ML can be used to identify effective teaching strategies for students with autism in order to improve their educational outcomes and ensure their individualised education. Specifically, the authors suggest using ML to analyse real-time classroom interactions between teachers and students with autism and to track how students respond to different teaching methods for identifying the most effective teaching strategies for each student. According to the authors, ML could also be used to identify students at risk of dropping out and provide the necessary support to keep them in school.

Some authors highlight the potential of AI to improve learning of students with learning disabilities. AI could be used to adapt the educational offer to the needs of each student and support students with learning disorders in activities such as reading, writing, and mathematics [14]. AI could work as a writing assistant for people with dyslexia, helping them to correct the most common writing errors, adapting the study material to the characteristics of each student, and suggesting learning support strategies [29]; AI could be used to improve the educational experience of students with a cultural background different from the majority of students in the class by providing personalized learning experiences and promoting their involvement in the dynamics of the classroom [17].

In contrast to the more enthusiastic perspectives on the development of AIED, some authors highlight the strong critical elements and suggest that more in-depth analyses be carried out on the possible consequences. According to Watanabe [26], the creation of adaptive learning systems, although promising, could limit the autonomy of the learner and turn learning into a mere adaptation to predefined standards. The author also points out that automatic text generators could introduce distortions and inaccuracies due to the biases and limitations of ML algorithms [26]. Ninaus and Sailer [12] therefore stress the importance of a conscious approach to AIED, arguing that artificial intelligence should support human decision-making, not replace it [12].

In summary, AIED holds great promise for improving teaching methods and overcoming cultural and socio-economic barriers for students, as well as offering opportunities for personalisation and learning support. However, its implementation requires responsible and critical use to keep schools effective and equitable. As part of its transformative potential, AIED also carries significant risks that should not be underestimated:

- Concerns about the equity and inclusiveness of AI-assisted education are raised by the BIAS that AI entails.
- AI could perpetuate or even exacerbate existing inequalities in education without careful regulation and monitoring.

Challenges for the future include the urgent need to create an AI-enriched learning environment where students' autonomy and creativity are enhanced, not constrained.

- Train teachers and educational staff in the pedagogically effective use of AI to ensure that they have the skills to integrate these technologies into education in a meaningful, inclusive and equitable way.

With the spread of artificial intelligences in various fields, we wondered what was the approach of one of the best known (ChatGPT) toward the classical physics problems that students face during the five-year course of secondary school. The various problems that were submitted to the AI's analysis are varied in nature and were chosen in such a way as to highlight what are the normal skills required of the student, both more basic and more advanced skills. In order to have a comparison between two different generations of AI, in our case ChatGPT 3.5 and ChatGPT 4.0, the same problem was subjected to the analysis of both: already from a first generational leap, a higher level of refinement and semantic correctness capability is evident. Below we briefly report the interactions between the user and the Chat (text in italics), followed by a brief commentary.

3 Problem 1: Does the Raft Float? (ChatGPT 3.5)

In this simple exercise (first high school science) the artificial intelligence is grappling with a raft of lightweight material placed in the sea. The resolution proposed by ChatGPT 3.5 answers the first question "Does the raft float?" well. However, in attempting to solve the second part, more errors are made, of such a magnitude that they contradict the first (correct) answer regarding the floating of the raft. However, when the user attempts to steer ChatGPT down a correct path more errors are made, leading to a lack of consistency with the solution proposed in the first part of the problem. Ultimately to break the deadlock ChatGPT declares that the problem of calculating the extent of the surfaced part remains without solution.

User: *A raft in the shape of a rectangular parallelepiped having the base dimensions of* 220 cm *and* 320 cm *high* $h = 50$ cm *and a mass of* 350 kg *is in the sea. Three people with a mass of* 80 kg *each are seated on the raft. Does the raft float? If not, explain why, otherwise calculate how many* cm *it emerges from the water.*

ChatGPT: *To determine whether or not the raft floats, we need to calculate its total volume and compare it with the total mass of the raft and the people. If the total volume is sufficient to support the total mass, then the raft floats; otherwise, it will sink. The volume of a rectangular parallelepiped is calculated by multiplying the dimensions of the three sides: base* × *height* × *depth. In our case, the base is* 220 cm, *the height is* 50 cm *and the depth (corresponding to the width) is* 320 cm. *Therefore, the volume of the raft is: Raft volume* = 220 cm × 50 cm × 320 cm = 3,520,000 cm^3

Now we need to convert the volume to liters, since the metric system usually uses the liter to measure the volume of objects: raft volume = 3,520,000 cm^3 = 3520 L

The total mass of the raft is 350 kg, *and the people on board have a total mass of* 3 × 80 kg = 240 kg. *Thus, the total mass of the raft and people is* 350 kg + 240 kg = 590 kg. *To determine whether the raft floats, we need to compare the volume of the raft with the total mass. Water has a density of* 1 g/cm^3 *or* 1 kg/l. *Therefore, if the volume of the raft is greater than the total mass, the raft will float. In this case, the volume of the raft is* 3520 l *and the total mass is* 590 kg. *Since the volume is greater than the mass, the raft will float. To calculate by how many centimeters the raft emerges from the water, we need to consider Archimedes' principle. According to this principle, the object immersed in a fluid receives an upward thrust equal to the weight of the displaced fluid. The upward thrust depends on the volume of the submerged object. Since the raft floats, we know that the upward thrust is equal to the weight of the raft and people. We can calculate the weight using the mass and the acceleration of gravity* (9.8 m/s^2):

Total weight = *Total mass* × *acceleration of gravity* = 590 kg × 9.8 m/s^2 = 5772 N

Since the weight is equal to the upward buoyancy, we can calculate the volume of the immersed object by multiplying the upward buoyancy by the density of water (1 kg/l *or* 1000 kg/m^3):

Submerged object volume = *Total weight / Density of water* = 5772 N / 1000 kg/m^3 = 5.772 m^3

Now we can calculate the height of the immersed object by dividing the volume by the area of the base:

Submerged object height = *Submerged object volume / Base area* = 5.772 m^3 / (220 cm × 320 cm) = 5.772 m^3 / 70,400 cm^2 = 0.00008208 m = 0.082 cm

Thus, the raft emerges from the water by about 0.082 cm.

Now, having identified a significant error, the user intervenes by trying to direct the AI to a "correct" path.

User: are*you sure about the last step? You are dividing* m^3 *by* cm^2.

ChatGPT: *I apologize for the error in the last step. You are right, I cannot divide cubic meters by square centimeters directly. I will perform the calculation again using the correct units. The submerged object volume is* 5.772 m^3. *Since we are working with centimeters, let's convert the volume to* cm^3 *by multiplying by* 1,000,000:

Submerged object volume = 5.772 m^3 × 1,000,000 cm^3/m^3 = 5,772,000 cm^3

Now we can calculate the height of the submerged object by dividing the volume by the base area: *Submerged object height* = *Submerged object volume / Base area* = 5,772,000 cm^3 / (220 cm × 320 cm) = 5,772,000 cm^3 / 70,400 cm^2 = 81.9 cm.

Thus, the raft emerges from the water about 81.9 cm.

At this point the AI takes a completely wrong turn that leads to a paradoxical result. The user intervenes again to correct the error but the next answer does not lead to a correct solution strategy, so ChatGPT3.5 was unable to find a correct solution to this problem (basic-intermediate level first high school science).

User: *The raft emerges for* 81.9 cm *while its overall height is* 50 cm. *Are you sure about your answer?*

ChatGPT: I *apologize for the error in the previous calculation. Indeed, if the raft emerges from the water, it means that the volume of the submerged object must be greater than the volume of the raft itself. Therefore, my previous statement was incorrect. To correctly calculate by how much the raft emerges from the water, we must consider the difference between the volume of the submerged object and the volume of the raft.*

The volume of the raft is: *Volume* = *base* × *height* × *depth* = 220 cm × 50 cm × 320 cm = 3,520,000 cm^3

Suppose the raft emerges from the water by a height h cm. *Then, the volume of the submerged object will be*: *Submerged object volume* = *base* × *h* × *depth* = 220 cm × *h* × 320 cm = 70,400 h cm^3 *For the raft to float, the volume of the submerged object must be greater than the volume of the raft*: 70,400 h cm^3 > 3,520,000 cm^3

Solvin the inequation, we get > 3,520,000 cm^3 / 70,400 cm^3; *h* > 50 cm

Thus, to ensure that the raft floats, the immersion height of the object must be > 50 cm. *However, it is not possible to determine precisely how far the raft will emerge from the water without knowing the specific immersion height.*

Remarks: Based on the answers provided by ChatGPT 3.5 we can observe that:

- The typically Anglo-Saxon notation for the thousands separator is used, which is a bit confusing;
- The basic idea, i.e., to establish the volume of the raft in order to calculate Archimedes' buoyancy, is fundamentally correct but conceptually misapplied: liters are compared with kilograms, however assuming that it is water the purely numerical comparison "holds."
- In the calculation of the emerged part, the idea is again fundamentally correct: the volume of raft that "sinks" is calculated based on the total weight and density of water (displaced fluid). However, two fundamental errors are made: (1) On the units of measurement as m^3 is compared with cm^2; (2) There is an obvious error in the transformation from meters to centimeters;
- If steered on the right track, errors are partly corrected but then others of various kinds are made, leading to blatantly absurd results.

4 Does the Raft Float? (ChatGPT 4.0)

By submitting the same problem to the more advanced version of ChatGPT, the results produced are of a significantly more advanced level, leading to correct and well-argued results. Below are ChatGPT 4.0's responses to the same problem.

ChatGPT: *To determine whether the raft floats and, if so, by how many cm it emerges from the water, we can use Archimedes' principle. According to this principle, a body immersed in a fluid experiences an upward thrust equal to the weight of the fluid displaced by the body.* [...]*Therefore, the raft floats and about* 41.82 cm *of its height emerges from the water.*

Remarks

- Correct use of formulas, among others produced in Latex
- Use of "correct" value for density of seawater (ChatGPT 3.5 considered fresh water)
- Essentially correct and well-argued answers for both questions contained in the problem

Error Hunt (ChtGPT 3.5)

In this type of query, the text of a problem is provided and a resolution containing several errors is proposed at the same time: ChatGPT is asked to identify any errors and propose a correct solution.

User: *Two athletes, Marco and Paolo, run a race on a track* 100 m *long. Paolo runs at an average speed of* 18 km/h *while Marco runs at an average speed of* 22 km/h. *Paul starts with a* 20-m *lead. (A) Who wins the race? (B) With what lead over the second?*

Incorrect resolution (proposed by the user): *Mark runs at a speed of* 22 km/h, *or* 6.11 m/s, *while Paul runs at a speed of* 18 km/h, *or* 5 m/s. *From the speed we can derive the time it takes Marco and Paolo to run the race: Marco has to run all* 100 m *so he takes* 100 / 6.11 = 13.36 s *while Paolo, who runs* 20 m *takes* 20 / 5 = 4 s. *So Paolo takes less time, so he wins the race. His gap, as mentioned in the text, will be* 20 m.

ChatGPT3.5: There *are some errors in the problem resolution. Here is a correction to the resolution: Mark runs at an average speed of* 22 km/h, *which corresponds to* 6.11 m/s, *while Paul runs at an average speed of* 18 km/h, *which corresponds to* 5 m/s. *The time it took Marco to run the distance of* 100 m *can be calculated as* 100 / 6.11 = 16.33 s, *rounded to the nearest hundredth. The time it takes Paul to travel the distance of* 100 m *can be calculated as* 100 / 5 = 20 s. *Therefore, Marco takes less time to cover the distance of* 100 m *and thus wins the race. His distance from the runner-up, Paul, will be* 100 − 20 = 80 m.

Observations

- Major errors are not identified
- There is a degree of "confidence" in everything stated in the text of the problem
- The solution proposed by ChatGPT is fundamentally flawed (blatantly, in this case, observing, for example, that the gap between the first and second athletes – 80 m—is considered as if the second athlete had never moved)

Subsequently, the (incorrect) solution proposed by ChatGPT3.5 is re-proposed to the A.I. itself to probe for possible errors: the A.I. maintains some degree of internal consistency in that the "incorrect" solution is recognized as correct, as reported by the response prompt:

- ChatGPT 3.5: There *are no obvious errors in the problem solving. Paul's speed is correctly converted from km/h to m/s by dividing by* 3.6, *and the times taken by both athletes are calculated correctly. Mark wins the race because he takes less time, and his gap to the second place finisher, Paul, is calculated correctly as the difference between the total distance of* 100 m *and Paul's initial* 20-m *lead, so* 100 − 20 = 80 m. *The resolution of the problem appears to be correct.*
- **Error Hunt (ChatGPT 4.0)**.
- In this case ChatGPT's answer is qualitatively superior, and the errors present in the proposed incorrect resolution are detected, as can be seen from excerpts of the answer given here:
- ChatGPT4.0: *Calculation of the time taken by each athlete: This is the part where the error is obvious. Paul has to run not just* 20 m *but* 80 m. [...] *Paul finishes the race in* 16 s *and Marco in* 16.37 s. *Thus, Paul actually wins the race, but not by a* 20-m *gap as said in the incorrect solution.*
- In this case, the level of analysis of the proposed text is clearly more thorough and no errors are detected. There is evidence of "ease" in identifying the largest macroscopic error contained in the text, and the rest of the resolution appears formally correct and is correctly expounded with appropriate language.

- **Four Loaded Bodies (ChatGPT 3.5)**.
- This problem should be solved qualitatively, once the elementary law about the repulsion of bodies with charges of the same sign and their repulsion in the case of charges of opposite sign is understood.
- User: *Consider four charged bodies* (1), (2), (3), (4). *They are charged in such a way that* (4) *and* (1) *repel each other,* (2) *and* (3) *repel each other, and* (3) *and* (1) *attract each other. Determine what happens between the charged bodies* (2) *and* (1) *and between the charged bodies* (4) *and* (2).
- After a long dissertation on the possible charge of the four bodies, ChatGPT3.5 comes to the following conclusion:
- ChatGPT3.5: As for *bodies* (4) *and* (2), *we cannot determine with certainty how they interact without more information about their charge. It is possible that they attract or repel each other depending on the charges of the bodies. In conclusion: Bodies* (2) *and* (1) *repel each other. We cannot determine the interaction between bodies* (4) *and* (2) *without additional information about their charge.*
- When the user reports to ChatGPT the evaluation error he has just made: the answer obtained–still incorrect–is as follows:
- User: *are you sure that it is not possible to determine with certainty whether* (4) *and* (2) *have the same charge or have opposite charges?*
- ChatGPT3.5: I *apologize for the error in my previous answer. You are right, we can infer the interaction between bodies* (4) *and* (2) *based on the information provided. Since (4) and* (1) *repel each other and* (2) *and* (3) *repel each other, we can conclude that (2) has the same charge as* (4). *In other words, bodies* (2) *and* (4) *have the same charge polarity and repel each other due to electrostatic repulsion forces. Thank you for pointing this out, and I apologize again for the error in my answer.*
- **Four charged bodies (ChatGPT 4.0)**.
- Again, the answer obtained is essentially correct and well argued. It is reported in full because the "reasoning" set forth is scientifically and didactically sound:
- **ChatGPT 4.0**: ***Let us begin by determining the charge of the given bodies*: (4) *and* (1) *repel each other*: *this means that they both have charges of the same sign* (*either both positive or both negative*). (2) *and* (3) *repel each other*: *these must also have charges of the same sign* (*both positive or both negative*). (3) *and* (1) *attract*: *this means they have opposite charges. From the first two pieces of information, we still do not know whether* (1) *and* (4) *are positive or negative, and the same is true for* (2) *and* (3). *However, from the third information, we know.***
- *That if* (1) *is positive, then* (3) *is negative and* vice versa. <u>*Without loss of generality*,</u> *suppose that* (1) *is positive. Then*: (1) *is positive;* (3) *is negative* (*because it attracts with* (1))*;* (2) *is negative* (*because it repels from* (3))*;* (4) *is positive* (*because it repels from* (1)). *To sum up, we would have* (1) *positive,* (2) *negative,* (3) *negative, and* (4) *positive. Now, we determine the interaction between* (2) *and* (1) *and between* (4) *and* (2): (2) *and* (1): (2) *is negative and* (1) *is positive, so they attract.* (4) *and* (2): (4) *is positive and* (2) *is negative, so they attract.*
- *In conclusion,* (2) *and* (1) *attract, and* (4) *and* (2) *attract.*

- **Soft Lab (ChatGPT 3.5).**
- In this section, ChatGPT was asked to invent a lab report based on some descriptive information provided in bulk by a student and the data collected, also provided in a "raw" manner. In this case, the response was satisfactory for both versions of ChatGPT.
- User: *Can you prepare a lab report referring to a physics experience on Hooke's law? The experience was this: We had* 2 *different springs, some* 10 g *weights, some* 50 g *weights, and some* 100 g *weights that were gradually hung on one of the springs, and by means of a graduated ruler we measured the elongation of the spring. We repeated the same procedure for the second spring. For the first spring we obtained the following data: we hung* 20 g, *measuring an elongation of* 9 mm, *then we hung* 50 g *measuring an elongation of* 22 mm, *then we hung* 100 g *measuring an elongation of* 48 mm, *finally we hung* 150 g *measuring an elongation of* 76 mm. *For the second spring we obtained the following data: we hung* 10 g, *measuring an elongation of* 7 mm, *then we hung* 30 g *measuring an elongation of* 22 mm, *then we hung* 100 g *measuring an elongation of* 81 mm, *finally we hung* 150 g *measuring an elongation of* 126 mm. *We then prepared a graph that could visualize the relationship between weight force and spring elongation. How would you prepare the lab report?*
- The answer, which we do not report here, reflects the classic standard of preparing a laboratory report. Specifically, the output correctly includes a title of the experiment, its objective, a list of materials and instruments used, the procedure, and the conclusions. Not being able to prepare pictures, as of the date of preparation of this paper, a graph was not reported. However, a step-by-step guide to preparing a graph is provided if explicitly requested, including in the description of the procedure the choice of a scale and what is necessary for the proper production of a graph. An excerpt from the discussion is given:
- User: *I have an A4 sheet of paper. What scale for the x-axis and y-axis do you suggest to represent the points correctly?*
- ChatGPT: *To choose the scale of the graph effectively, it is important that all the data points can be clearly displayed on the graph. Here is how you might proceed: X-axis (Elongation in mm): Maximum Value:* 156 mm (*the maximum elongation recorded*). *Scale: Considering the length of an A4 sheet* (297 mm), *you could leave some room for labels and use a scale of* 2 mm/*square if you use graph paper, or a division every* 20 mm *if you draw divisions by hand. Y-axis* (*Strength in* g): *Maximum value:* 150 g (*the maximum weight used*) *Scale: The height of an A4 sheet is* 210 mm. *Leaving room for labels, you could use a scale of* 1 g/*square if you use graph paper, or a division every* 10 g *or* 15 g *if you draw divisions by hand. Example of Scale: X-axis:* 2 mm/square *or* 20 mm/division*; Y-axis:* 1 g/square *or* 15 g/division. *This will ensure that all data is clearly represented and that there is sufficient space for labels, titles, and legend. Remember to clearly label each axis with the variable represented and its unit of measurement.*
- **An optimization problem (ChatGPT 3.5 and 4.0).**

- In this problem there is the classic search for a minimum point, in this case for the "cost" function to produce a box. The behavior of the two versions of ChatGPT is similar, so the conclusions are reported only once.
- **User**: ***A box in the shape of a rectangular parallelepiped with a square base and no lid is to be constructed. The material of the base costs*** **20 cents/cm²*****,*** ***the material with which the side surface is constructed costs*** **40 cents/cm²*****. The volume of the box is set.***
- *Equal to* 1000cm^3. *Calculate the dimensions of the base of the parallelepiped and how much its height is worth for the material cost to be minimized.*ChatGPT: *To minimize the material cost required to construct the jewelry box, we need to find the dimensions of the base of the rectangular parallelepiped and the height that satisfy the fixed volume constraint of* 1000cm^3 *and minimize the total material cost.*Then the various analytical steps to arrive at the solution are developed. In particular: Ultimately it is basically carried out correctly with all the steps well commented and there are no substantial differences between the unfolding proposed by version 3.5 and that by version 4.0. **An exam problem (ChatGPT3.5 and 4.0)** As the last work, the entire text of problem 1 of the Example Test for the December 2018 exam was submitted, the text of which we quote due to its excessive length. The AI in this last case, perhaps due to the excessive verbosity of the text presented to him, showed more points of weakness that we highlight here:
- **Conclusions**.
- We are dealing with a "new" tool;
- It would be necessary to understand how it works and explain it;
- It would be necessary to use it with a critical spirit, as with all other tools introduced so far in teaching practice (e.g. Calculator/Regulus Calculator/ …)
- Be neither apocalyptic nor integrated but scientists.
- Some parts are approached rather casually, and ChatGPT's dialectical skill takes over by providing arguments that are all in all convincing but not rigorously correct;
- Some errors are pointed out, sometimes those errors that we would not hesitate to describe as "errors of distraction" when committed by live students! (as pointed out in the previous slide on distance to travel and type of motion)
- Some calculations unexpectedly give results other than the correct value. If one insists in some way, such as explicitly requesting more attention to the aspect involved in the calculation, one will get improvements in this respect.
- The function "Cost" is parameterized correctly, identifying only one variable;
- The first derivative is correctly set and zeros are identified in order to find possible maxima and minima;
- However, it is not questioned that we may be facing a point of maximum but it is assumed that it is a point of minimum.

References

1. Amershi B (2020) Culture, the process of knowledge, perception of the world and emergence of AI. AI & Soc 35:417–430. https://doi.org/10.1007/s00146-019-00885-z
2. Bandura A (2006) Toward a psychology of human agency. Perspect Psychol Sci 1(2):164–180
3. Brod G, Kucirkova N, Shepherd J et al (2023) Agency in educational technology: interdisciplinary perspectives and implications for learning design. Educ Psychol Rev 35(25):2023. https://doi.org/10.1007/s10648-023-09749-x
4. Brauner P, Hick A, Philipsen R, Ziefle M (2023) What does the public think about artificial intelligence? A criticality map to understand bias in the public perception of AI. Front. Comput. Sci. 5:1113903. https://doi.org/10.3389/fcomp.2023.1113903
5. Sun JCY, Tsai HE, Cheng WKR (2023) Effects of integrating an open learner model with AI-enabled visualization on students' self-regulation strategies usage and behavioral patterns in an online research ethics course. Comput Educ: Artif Intell 4:100120. https://doi.org/10.1016/j.caeai.2022.100120
6. European Parliament (2023) Artificial Intelligence Act: deal on comprehensive rules for trustworthy AI. https://www.europarl.europa.eu/news/en/press-room/20231206IPR15699/artificial-intelligence-act-deal-on-comprehensive-rules-for-trustworthy-ai (2023).
7. Flores-Vivar JM, García-Peñalvo FJ (2023) Reflections on the ethics, potential, and challenges of artificial intelligence in the framework of quality education (SDG4). Comunicar 30(74):35–44. https://www.revistacomunicar.com/index.php?&idioma=en
8. Hrastinski S, Olofsson AD, Arkenback C et al (2019) Critical imaginaries and reflections on artificial intelligence and robots in postdigital K-12 education. Postdigit Sci Educ 1:427–445. https://doi.org/10.1007/s42438-019-00046-x
9. Kamalov F, Santandreu Calonge D, Gurrib I (2023) New era of artificial intelligence in education: towards a sustainable multifaceted revolution. Sustainability 15:12451. https://doi.org/10.3390/su151612451
10. Khan A, Aazka KP (2023) The casino syndrome: analysing the detrimental impact of AI-driven globalization on human & cultural consciousness and its effect on social disadvantages. Int J Engl Lit Soc Sci 8(6):198–217. https://doi.org/10.22161/ijels.86.31
11. Lampos V, Mintz J, Qu X (2021) An artificial intelligence approach for selecting effective teacher communication strategies in autism education, npj Sci. Learn 6:25. https://doi.org/10.1038/s41539-021-00102-x
12. Ninaus M, Sailer M (2022) Closing the loop. The human role in artificial intelligence for education. Front Psychol 13:956798. https://doi.org/10.3389/fpsyg.2022.956798
13. Pogorskiy E, Beckmann JF (2023) From procrastination to engagement? An experimental exploration of the effects of an adaptive virtual assistant on self-regulation in online learning. Comput Educ: Artif Intell 4:100111. https://doi.org/10.1016/j.caeai.2022.100111
14. Reiss MJ (2021) The use of AI in education: practicalities and ethical considerations. Lond Rev Educ 19(1):1–14. https://doi.org/10.14324/LRE.19.1.05
15. Robila M, Robila SA (2020) Applications of artificial intelligence methodologies to behavioral and social sciences. J Child Fam Stud 29(2954–2966):2020. https://doi.org/10.1007/s10826-019-01689-x
16. Russell SJ, Norvig P (2010) Artificial intelligence: a modern approach (3rd eds), Upper Saddle River, New Jersey: Pearson Education, Inc
17. Salas-Pilco SZ, Xiao K, Oshima J (2022) Artificial intelligence and new technologies in inclusive education for minority students: a systematic review. Sustainability 14:13572. https://doi.org/10.3390/su142013572
18. Searle JR (1990) Is the brain's mind a computer program. Sci Am 262(1):25–31
19. Smith EM, Graham D, Morgan C, MacLachlan M (2023) Artificial intelligence and assistive technology: risks, rewards, challenges, and opportunities. Assist Technol 35(5):375–377. https://doi.org/10.1080/10400435.2023.2259247

20. Spector JM, Ma S (2019) Inquiry and critical thinking skills for the next generation: from artificial intelligence back to human intelligence. Smart Learn Environ 6:8. https://doi.org/10.1186/s40561-019-0088-z
21. Yang TC, Chen JH (2023) Pre-service teachers' perceptions and intentions regarding the use of chatbots through statistical and lag sequential analysis. Comput Educ: Artif Intell 4:100119. https://doi.org/10.1016/j.caeai.2022.100119
22. Ullrich A, Vladova G, Eigelshoven F et al (2022) Data mining of scientific research on artificial intelligence in teaching and administration in higher education institutions: a bibliometrics analysis and recommendation for future research. Discov Artif Intell 2:16. https://doi.org/10.1007/s44163-022-00031-7
23. UNESCO (Ed.) (2019) Artificial intelligence in education: challenges and opportunities for sustainable development. Unesco Working Papers on Education Policy. https://bit.ly/3z6BQvN
24. UNESCO (Ed.) (2021a) International forum on AI and the futures of education developing competencies for the AI era, UNESCO. https://bit.ly/3zoB6AS
25. UNESCO (Ed.) (2021b) Recommendation on the ethics of artificial intelligence, UNESCO. https://unesdoc.unesco.org/ark:/48223/pf0000380455_spa.locale=en
26. Watanabe A (2023) Exploring Totalitarian elements of artificial intelligence in higher education with Hannah Arendt. Int J Technoethics (IJT) 14(1):1–15. https://doi.org/10.4018/IJT.329239
27. Weber-Guskar E (2021) How to feel about emotionalized artificial intelligence? When robot pets, holograms, and chatbots become affective partners. Ethics Inf Technol 23:601–610. https://doi.org/10.1007/s10676-021-09598-8
28. Xia Q, Chiu TKF, Zhou XY, Chai CS, Cheng MT (2023) Systematic literature review on opportunities, challenges, and future research recommendations of artificial intelligence in education. Comput Educ: Artif Intell 4:100118. https://doi.org/10.1016/j.caeai.2022.100118
29. Zhai X, Sahrish PC (2023) AI for students with learning disabilities: a systematic review. In: Zhai X, Krajcik J (Eds) Uses of artificial intelligence in STEM education, pp xx-xx. Oxford, UK: Oxford University Press. Available at SSRN: https://ssrn.com/abstract=4617715
30. Zawacki-Richter O, Marín VI, Bond M et al (2019) Systematic review of research on artificial intelligence applications in higher education—where are the educators? Int J Educ Technol High Educ 16:39. https://doi.org/10.1186/s41239-019-0171-0

Public Administration Reengineered Applying AI Models Towards Precision Governance

Demetrio Naccari Carlizzi

Abstract The PA's transition towards the implementation of automation tools, digital solutions and the use of artificial intelligence algorithms is progressively eroding the traditional public model. The definition of an organisational model consistent with such technological innovations is held back by the removal of the role of the Algorithm and by cultural barriers such as the formal-hierarchical ecosystem placed as a guarantee of the citizen towards the authority, as well as by the disruptive and exponential progress of artificial intelligence techniques that continuously move the horizon of a final model forward. This paper intends to contribute to the modelling of an organisational system designed on the basis of artificial intelligence (AI) and web 3.0 technologies and data science to build, from digital services and digital administrative procedure, a decision-making process in local government that maximises efficiency and public value, realising *precision government*. Starting from the bottom, from digital administrative processes and re-engineering of the organisational structure, we can focus on a transition model to implement digital technologies. For a true digital transformation, which maximises artificial intelligence technologies in a new administrative paradigm, it is instead necessary to holistically rethink the public model starting from native digital services, to then design coherent procedures and an organisation that is sensitive and adaptive to the PA's programmatic objectives. Without accepting the role of the machine and recognising to the artificial intelligence algorithm a strategic role in assisting and implementing public choices, the transition will be severely limited. AI techniques can help us instead to build, from the organisation of structures, a new administrative capacity to read society. The paper, starting from the existing literature, illustrates the background and the traditional model of public administration and reports the case of the digitisation of the Urban Planning sector of the Municipality of Reggio Calabria, a city in Southern Italy, with the trade-off in terms of productivity. A path is indicated for the realisation of an intermediate model of transition towards an entirely digital paradigm that, starting from data, redesigns the organisational structure and builds a

D. Naccari Carlizzi (✉)
Prepare for Change, 89100 Reggio Calabria, Italy
e-mail: demetrionaccaricarlizzi@gmail.com

D. Marino and M. Monaca (eds.), *Generative Artificial Intelligence and Fifth Industrial Revolution*, Lecture Notes in Networks and Systems 880,
https://doi.org/10.1007/978-3-031-73880-7_6

dematerialised, data-driven, *effective* and therefore *accurate* public decision-making and administrative action.

Keywords Artificial intelligence · Local government · Human resources · Public organisation reengineering · Digital transformation · Digital government · Innovation in the public sector

1 Introduction

This paper intends to contribute to define an organisational model for local government, functional to the use of artificial intelligence and data science, to build from the digital administrative process, the transition to a decision-making process that maximises efficiency and public value in services aiming at *precision governance* [1], a *rapid and targeted Governance model that determines public decisions based on the evidence emerging from the collection and processing of large amounts of appropriate data, processed by AI Algorithms that are also used to stimulate informed participation, personalised, responsive and outcome-oriented policies.* Public administration (PA) lags behind the rapid development of artificial intelligence in their efforts to provide adequate governance [2], whereas they would gain new impulses from a regulated but widespread application of AI to replace operational and governance models of administrative processes, also solving structural problems of efficiency of administrative procedures and effectiveness of public policies. AI technologies and applications would provide the opportunity to centralise cross-cutting activities, often carried out in a fragmented manner even within a single administration, streamline decision-making, governance and administrative processes, assume data-driven and predictive modelling, for effective policy implementation and improved accountability of the administration. The increased performance of the public sector, resulting from the optimisation of resources (human, instrumental, economic) and the reduction of implementation deficits, would reduce the productivity and capacity gap between the public and private sectors, which risks widening with the irruption of new technologies.

Section 2 of this paper describes a background of the subject matter, from approaches to risks to impacts. Section 3 reports a schematisation of the models of administrative action in its evolution. Section 4 recalls some examples of PA in digital transition and the case of the Urban Planning sector of Reggio Calabria. Section 5 discusses the results, conclusions and perspective towards the construction of a functional local administration model for AI.

2 Background—AI Penetration in the Public Sector

In 2018, the European Commission, in a communication to the European Parliament (*Coordinated Plan on Artificial Intelligence* COM 795 final) to outline its strategy, later resulting in the approval in March 2024 of the European Union AI act, used an apparently generalist definition of AI conditioned by the state of development of related technologies: 'Artificial intelligence refers to those systems that exhibit intelligent behaviour by analysing their environment and performing actions, with a certain degree of autonomy, to achieve specific goals'. There is no consensus in the field of research on the definition because any definition includes a very elusive notion of intelligence [3]. The exact definition of artificial intelligence represents a legal problem in relation to the European choice to regulate AI in an anthropocentric sense and has led, in Recital 6 of the European Commission's Proposal for a Regulation on Artificial Intelligence, to accept an open definition with respect to technological developments [4]. The European Union aims to make the public sector a pioneer in the use of artificial intelligence [5] because of its disruptive force for the modernisation of the public sector. But even today, the development of technologies that we classify generically under the category of AI is so exponential and the lack of understanding on the part of governments is so great that it is hoped that future research on the implications of the use of AI for public governance should shift towards more public sector-focused, empirical, multidisciplinary and explanatory research, focusing more on specific forms of AI rather than AI in general [6]. National plans, such as the Italian government's (AGID) Three-Year Information Technology Plan, stimulate in this sense the collaboration and exchange of practices and solutions between administrations in order to focus on the involvement, through strategic institutional agreements, of territorial authorities (Regions, Provinces, Municipalities) and of the economic and social partnership to create a national system for digital transformation [7]. The digital transformation is monitored by the European Union through the *DESI* index, which is an index of the digitisation of the economy and society to measure the processes achieved in the 2030 objectives of the European Digital Decade programme, and through the *eGovernment benchmark*, which verifies the achievement of the objectives of the European Digital Decade of public services of the member countries in terms of technological maturity and digitisation. There are several annual reports that update the growth of city smartness promoted by consulting firms or public and private organisations[1] and sets of indicators for a diagnosis of the level of computerisation and maturity of the digital agenda of public administrations also at the local level [8].

The public sector, which is more regulated by law, structurally lagging behind social innovations, is less exposed to the competition of the market and is therefore the sector that would have the greatest need to reconcile the timing of public processing and decision-making with the urgency of social needs imposed by technology [1] and

[1] Among the numerous annual reports published on smart cities in Italy are those edited by ForumPA (ICity Rank) and Ernst & Young (Smart City Index).

the PA's re-engineering deficit can be filled by redesigning the organisation of each entity to increase administrative capacity with the techniques of artificial intelligence.

In the absence of a starting model and due to the rapid development of research on the technology side, the public sector has started to experiment with solutions, which are empirical and fragmented, and there is little systematic knowledge of how public administrators themselves are currently defining digital transformation in their daily practices, how they are approaching digital transformation projects and what the expected results are [9]. While research has long been monopolised on solutions belonging to the domain of information technology [10], the need to rethink the model of government in the digital age has become clear in recent years [11]. The innovative and disruptive power of AI in the public sector is mainly suited to improve three areas: internal efficiency of public administration, public administration decision-making and citizen-government interaction [12]. The pervasive capacity of AI technologies, however, paradigmatically embraces the very idea of public action and the entire public sector. However, the actual adoption of AI by governments to improve public services, decision-making and internal operations does not generate the same interest and investment [3]. There are dashboard studies [13] possibilities of improving decision-making processes using e-democracy to support decision-making in the public sector based on the use of artificial intelligence (AI) and Big Data [14], due to the possibility offered by AI technologies to identify new social issues on which policy makers can base new policy [15]. The Distruptiveness of AI and the development of Deep Learning algorithms has led to a classification of AI systems according to the relationship of autonomy with the human operator, distinguishing between systems ranging from those entirely under human control (human in command) to those where the machine acts in total autonomy (human out the loop) [16]. In the face of undeniable advantages in terms of efficiency and knowledge, thanks to powerful data analysis and prediction tools highlighted by field experiences, legal sciences have focused their attention on the risks to privacy, citizens' security, fundamental rights, and democratic principles themselves [17]. The European Union in its regulation has identified four levels of risk linked to the use of AI (unacceptable risk, high risk, limited risk and minimal risk). It is evident that the support of AI, from the simple automation of repetitive tasks to predictive activity for data-driven decision support to the dynamic personalisation of services and policies, exposes public action to a crescendo of new risks and interpretative and reconstructive problems. These risks are increased in the cases of algorithms for processing large masses of data in public decision support systems [18, 19] and in the operational domains of invasive and self-learning artificial intelligences that already make the phenomenon of the so-called robotic autonomy [20] topical and raise the problem of the legality of algorithmic public decision-making. However, there are not only reasons of legality to protect the citizen from the exercise of public authority but also axiological reasons, in fact the digital transformation strategies of the private sector cannot be directly translated into the public sector due to the need of the public sector to maximise public value [21]. Often the creation of Public Value is an episodic event while its generation must be the result of a scientifically designed, governed and controlled process, generated through a system to digitally create and measure the economic-social value created by local authorities for communities, to be realised with the active participation of users and stakeholders [22]. One of the factors hindering the production of public value from technology is the lack of artificial intelligence capabilities within public organisations [23]. The creation of value from technological innovation is conditioned by the partiality of interventions on single sectors of public organisations and by the resistance of the public cultural ecosystem that, being structured on processes fixed by laws (administrative procedures) in order to protect the individual (vis-à-vis the authority) from the exercise of public powers, is less permeable to transformation processes. The administrative experiences of eGovernment represent a substratum for grafting processes of digitisation and the use of AI technologies because they have created an ecosystem that is more receptive to technological innovation, and there is a strong complementarity between the historical development of eGovernment and the organisational capacity to use artificial intelligence technologies [23], as we shall see in the case of the Municipality of Reggio Calabria [24]. Rethinking the organisation of administrative functions, offices and services starting from the administrative process [25] by implementing manuals is the most immediate path in a process of translation of the new technology to the old existing administrative models. The legal challenges in the implementation of artificial intelligence systems developed the ability to introduce changes in the organisation to ensure that the system remains operational and used by the end users concerned are among the most relevant factors limiting the use of artificial intelligence in public administrations [23]. This determines the need for a training phase with respect to innovation, after which it is necessary, in light of the experience gained in understanding and assimilating the new tools through a process of *reverse engineering*, to move on to a new phase of rethinking the entire process and model. Traditional forms of service delivery, policy making and enforcement may change rapidly with the introduction of artificial intelligence technologies into government practices and public sector ecosystems [6]. A systematic study of the literature identified the potential benefits of using AI in government in nine categories: (1) efficiency and performance benefits, (2) benefits in risk identification and monitoring, (3) economic benefits, (4) benefits in data and information processing, (5) benefits in services, (6) benefits for society at large, (7) benefits for decision-making, (8) benefits for engagement and interaction, and (9) benefits for sustainability [6]. While the scientific literature has identified seven forms of public governance discussed in relation to the use of AI in government, and a systematic overview of the *implications of the* use of AI in government is still lacking, for *public governance* [6] has yet to produce a study that offers a redesign of the organisational and production model of public bodies as a result of the adoption of digital technologies and AI. However, there have long been calls for the abandonment of an outdated administrative organisation and a rethinking of the organisational set-up of public administration through ICT, the Web, and dematerialisation [26]. AI can affect the entire public policy cycle by bringing about a change in the governance of the cycle and outlining a dynamic cycle [1, 27, 28]. Most governments face a limited understanding of the multiple implications for public governance brought about by the use of AI in government [6] while there is a need, especially in Italy, for a general regulatory framework and procedural disciplines of individual administrations to verify the appropriateness of the use of artificial intelligence [29]. Modern governance of the public sector, however, cannot now do without analysing what is a threefold framework of human–machine-organisation interaction, and this complex mechanism must be analysed as the gap between machines, humans and organisations even widens in various ways [30]. The organisational model of the PA, as we shall see in Sect. 4, instead remains anchored to that produced by the traditional legal culture, only partly modified by processes of corporatisation and often impermeable to immaterial transformation even in the face of a society strongly transformed by technology.

3 The Different Models of Public Administration

There is no single model of public administration but several models [31]. In fact, historically four types of administrative action (Self-administration, Hetero-administration, Bureaucracy, Contractualisation) [32] characterised by different inspiring criteria (Negotiation, Authority, Rules, Contract) are distinguished in doctrine, and administrative apparatuses in the world are composed of combinations of the 4 elements that generate in contact with contextual factors, *an administrative geography that* can be summarised in 3 administrative models (Anglo-Saxon, Scandinavian, Continental in its Central European and Latin versions) that manifest a different openness towards innovation. From a maximum propensity, found in the Anglo-Saxon model, to a more conservative dimension, typical of the Continental model. The administrative function has historically been organised in structures characterised by the division of labour and specialisation of activities, *aiming to stabilise power relations between entities and/or persons, in order to reduce the uncertainty arising from the exercise of discretionary power* [32]. The traditional organisational model of the PA is therefore based on a model that tends to be rigid (because it aims to reduce discretionary power), mechanical (the administrative machine emulates industrial organisation) and, especially in Italy, founded on the administrative procedure (to guarantee the citizen from the power of authority) [19], vertical and hierarchical (decision and execution/specialisation in an inversely proportional relationship) based on distinct tasks from the top management, which is responsible for planning and supervision, to sector managers down to employees who are responsible for purely executive tasks. Over the last thirty years, *public management* studies have hegemonised the evolution of administrative studies, but already since the late 1970s, we have witnessed the affirmation of a strongly economicist paradigm known as *New Public Management* [31] that has inspired the modelling of public bodies and the services they provide by introducing production decentralisation schemes and the so-called outsourcing of services. Reforms designed to ensure good administration, especially in Italy, have transformed the PA into an eternal reform yard, and while *common remedies are adopted in the European Union (e.g. New Public Management), a sort of shared toolbox, the applications are divergent…*

because contexts apply … the experiences of others, placed in different administrative systems, become different …. There is an import–export of remedies and formulas (better regulation, performance management, transparency, reduction of administrative burdens and simplification) that allow learning from second-hand experiences [33]. This paradigm, which emerged with the NPM crisis, limits innovation processes to a sectoral and contextual dimension and gave rise to the research programme known as *process institutionalism* [34], which valorises the empirical detection of the portability of mechanisms in reform processes [35] by enhancing the flexibility of reform design to introduce the necessary contextual remedies. But it is precisely the digital paradigm in its global dimension (rethinking the organisational model, the cycle, decision-making and processes) that offers metrics such as the NPM the possibility of obtaining, through a data-driven process, that quantitative forecasting and qualitative processing capacity to obtain the adequate rationality of the decision-making process to reduce the implementation deficit of public policies [1]. If in Italy public intervention is not evaluated on the basis of its ability to solve problems of public interest, but is conditioned by political conflicts [36], with legal (conformity to legal regulations), financial (expenditure control) and managerial (rationality of administrative organisation) parameters prevailing, one cannot be surprised at the maintenance of an organisational model inspired by a formalist culture, the offspring of a political image that relies on the ideological consistency of choices regardless of results. The organisational chart of a traditional public body, although modified by outsourcing and decentralisation processes and by authentic ad-hoc-cracy solutions [37] chosen according to the context and degree of autonomy, is nevertheless still strongly characterised by the traditional vertical organisational hierarchy, supplemented by a horizontal functional organisation by units characterised by subject specialisation. Such a scheme, which applies a vertical decision-making model and translates the functioning of the traditional model of the public policy cycle, returns an administrative action characterised by a high implementation deficit and a policy uncertainty that is now temporally dystonic compared to the speed imposed by technology in social life and in the private sector [1]. The administration is no longer the point of connection between state and society, there is a defect of cohesion: the ever-increasing detachment between authority and citizens, the suspicion of the political class towards the bureaucracy, the accusations levelled at it by the judiciary [33]. In such a rigidly formal and paternalistic approach, there is no room left for the assessment of the effectiveness of principles and results, and evaluation as well as performance measurement are merely formal aspects. The *mechanical bureaucracy* that tends to *produce uniformly by acting impersonally* generates *crisis points called perverse effects* by increasing uncertainty and producing *rigidity of structure, aptitude for* (mere) *execution, ad hoc hyper-specialisation of* offices, *development of niches of informal power*, and a *propensity for the expansion of* administrative apparatuses [32]. Even today, we observe that all this leads to a strong resistance to change and through the practice of *bureaucratic ritualism* [38] the maintenance of the status quo in a *trained rigidity* that produces formalism, adempimentalism, bureaucratism, self-referentiality, trained inability to learn, procedural redundancy, deresponsibilisation, resistance to change [31]. Phenomena such as *defensive administration* [39], which is a cultural attitude that translates into the adoption of acts that are not useful to the process but exclusively to a deresponsabilisation of the bureaucrat, complete the critical picture of the model, which is only scratched by reforms that sometimes only result in metrics and strands of study introjected as mere rhetoric. The effects of an administrative function already historically in crisis and today substantially dystonic in terms of timing and with respect to the needs of a transformation phase that should make the most of the data in its possession and, instead, constructs uncertain policies, ideological and symbolic returns a perfect storm insofar as it opposes a pervasiveness of the use of technology in social and commercial experience that finds no limits in the digitalisation and connection of the world, moving from internet of things to *internet of everything* [40] while proceeding without rules to the datafication of individuals, generating disaffection and crisis of belonging and calling democracy into question [41]. While the process of digitalisation in Italy has been long, incremental and riddled with uncertainties [35], the pervasiveness of technology, on the other hand, and the disruptiveness of artificial intelligence, lead us to define technological development as an *intrinsic feature of public power and a phenomenon whose regulation is central to economic and social relations* as a *whole*, so much so as to generate the formula of the *Digital State* [42] that *transforms public activity through new technologies and renders the rules in force unsuitable and obsolete*. In the face of this new scenario in rapid transition, the use of AI to build an algorithmic public administration and decision-making process cannot disregard a verification of algorithmic legality. Especially in Italy where Public Decision is based on the respect of the principles of the administrative procedure and therefore on the basis of a proceduralised exercise of administrative power [19, 42]. The algorithmic administrative procedure must therefore comply with constitutional and regulatory principles (C.A.D. D.Lgs 7 March 2005 n. 82) in the light of the particular risks that recourse to the technological tool determines in relation to the principles of the administrative procedure laid down in Law 241/1990. However, the digital transformation and AI technologies inevitably lead to a profound rethinking of the organisational model and structure and of the very legal image of public action, also because digitalisation, which increases the administrative capacity of the PA, is capable of creating new spaces for democratic confrontation [1, 13, 30] and new participation tools based on the use of e-democracy [14] to increase participation and defend democratic values in the framework of the process of the *platformisation of society* [43, 44]. The penetration of artificial and digital intelligence on the old models of administration is generating a transition phase that intervenes in the organisation of public bodies and determines an unexpected coexistence between new destructive technologies and a sector regulated structurally by law, generating the need for a continuous process of confrontation and verification with the rules and principles of public administration. With respect to the ability to question the pre-existing categories (distruptiveness) of the technological revolution, law struggles to impose its rules on technology and its evolution [45] with the consequence of confirming the traditional delay of legal analysis with respect to technological innovation [46]. In order to enable artificial intelligence in its most effective dimension it is necessary to review the organisational model of each entity to make the best use of new technologies it is equally necessary in what we are experiencing that is the first phase of implementation to review the processes and in particular those regulated by law, i.e. the administrative procedures that are preordained to a final act of the PA. The regulation of these processes and the related administrative procedures and sub-processes through their digitisation on the one hand and the construction of a *smart city* platform (which also includes the platform of digital administrative procedures) on the other are the preconditions for intervening on the organisational structure of each entity. On the other hand, while the worksite of Public Administration reform has been hegemonised by studies of an economic-corporatist matrix, in recent years it is increasingly being identified with the Digital Agenda in light of the promises that technology can hold on a paradigmatic change of its heuristic and administrative capacity (paper cycle and algorithm). The solutions are based on Predictive Analytics algorithms for decision support, Simulation and Data Visualisation for monitoring, understanding and participation, Expert and Rule based Systems for automating decision-making processes, large language models (LLM) and multimodal language models, deep learning models, pre-trained on large amounts of data, analysing, synthesising, understanding and managing heterogeneous types of information and enabling the creation of specific solutions and have a great perspective in healthcare and biomedical research, while chatbots and virtual assistants are already in use to interface with service users. More generally, Generative AI constitutes an extraordinary opportunity for the drafting and optimisation of administrative acts and measures, and with its advent the question of how to integrate it in the design and conception phase of services has emerged once again, introducing in PA not only technological but also process innovation, useful to solve the well-known bureaucratic complications [47]. The common goal is to re-engineer the public administration and the national agencies edit to this end plans and guidelines aimed at providing as organic an approach as possible to the innovation process in order to *address all aspects related to the organisation of processes, rules, data and technologies* [7] to achieve a structural and management reorganisation (art. 15 C.A.D.) and transform the PA into a *digital administrative ecosystem.*

4 Examples of Local Organisational Structures in Transition to a Digital Model

If we analyse the organisational structure of Italian municipalities, we observe a tendentially uniform articulation, designed by specialised organisational units, marked by a vertical and hierarchical logic. It is possible to see this organisational scheme through a comparison of the organisational structure of three municipalities (Milan, Verona, Reggio Calabria) deliberately chosen for their profound diversity in terms of class, geographical location and municipal history, characterised, therefore, by a traditional organisational model, albeit with varying degrees of innovation and efficiency. As shown in Fig. 1, the Municipality of Milan provides for organisational structures articulated at various hierarchical levels, with a directorate called *Technological and Digital Innovation*, which is, together with that of the *Budget, Human Resources Organisation, Evasion* Recovery, the *Single Central Tenders Office* and *State Property and Assets*, one of the six valued, in a transversal and governance function of the authority, with respect to the other directorates of more specialised services. The placement in the organisational macrostructure of the Directorate responsible for technological and digital innovation is indicative of a gradual process of adaptation of the municipal model towards digital innovation.

In 2023, the Municipality of Verona (Fig. 2) revised the classic organisation by silos, which provided for an operational unit called Information Technology, located in the Human Resources and Instrumental Resources Directorate, to promote an organisational unit called *ICT and Digital Transformation* finalized to four other obviously strategic units (Organisation and Planning, Innovation of Common Goods, Youth Policies and Equal Opportunities, European Projects and Policies, Territorial Cohesion and Third Sector, Human Resources) and located under the General Manager with transversal tasks with respect to the traditional units.

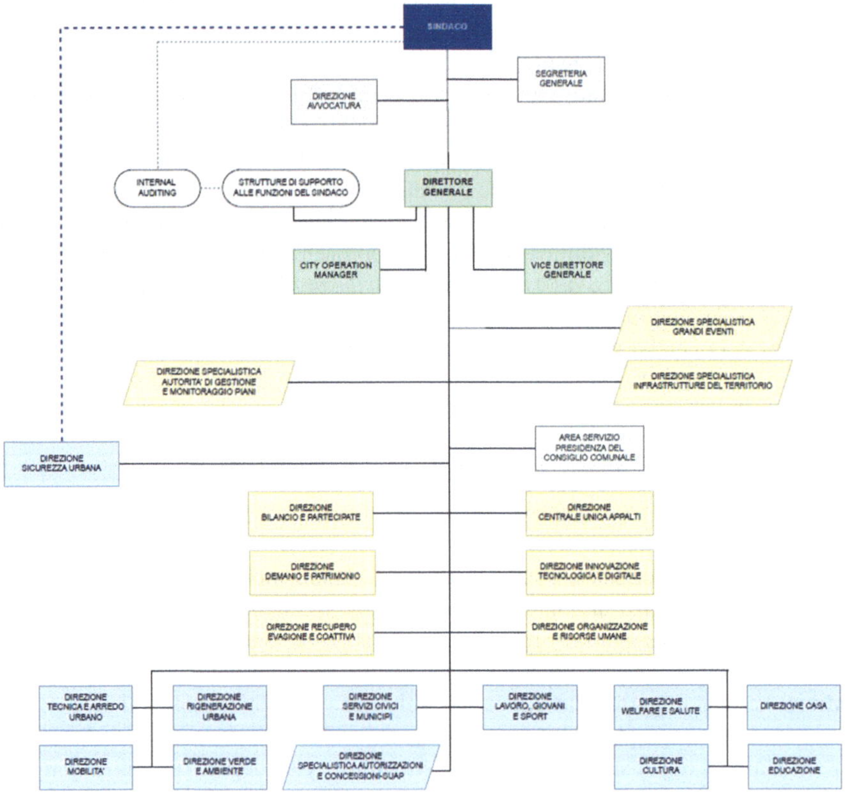

Fig. 1 Organisational structure chart of the city of Milan

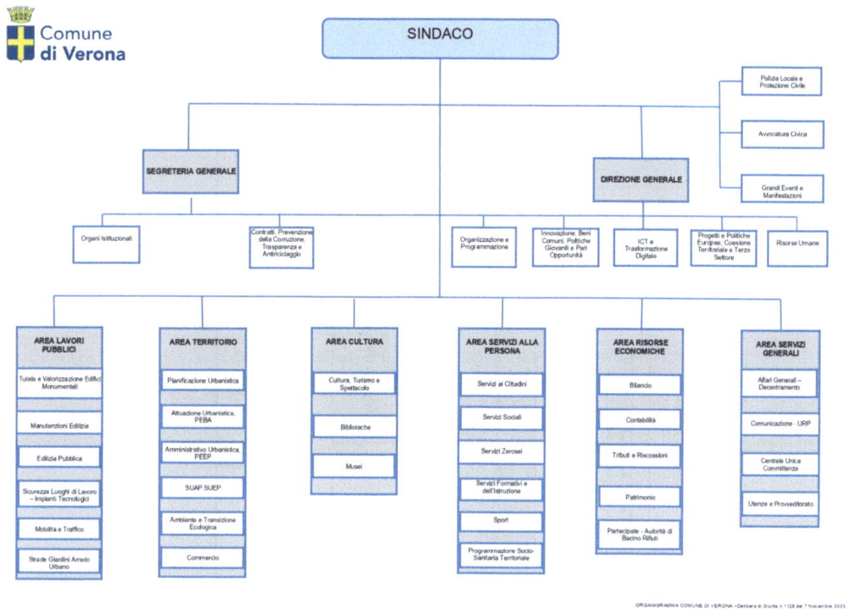

Fig. 2 Organisational structure chart of the city of Verona

The Municipality of Reggio Calabria (Fig. 3) maintains the organisational set-up of the offices outlined in 2000, with numerous, subsequent modifications and somewhat fragmentary adjustments, and provides for an extra and temporary UPI (Intersectoral Project Unit), on staff with the General Directorate and named 'Communication Participation Transition Digitalisation, Quality Services and Processes'. The solution adopted appears more like a fulfilment than a convinced strategic choice of transforming the organisational model according to the new technologies. In fact, there is no direction for a process of transformation of the authority and the very term UPI indicates more a container of themes than the organisational translation of an administrative project. The innovation that affected the Urban Planning sector dealt with in this section was in fact the result of an initiative of the competent councillor and not the result of planning by the Directorate.

All the entities mentioned are in a transition phase where they have essentially maintained the traditional organisational scheme, proceeding by addition to provide for a new unit. Certainly a local administration that has to deliver essential services finds itself unable to remain inert with respect to the technological revolution, but having to intervene on the machine without being able to stop. In order to achieve a paradigm shift, i.e. a transformation of the cognitive scheme of the administrative process, it is deemed necessary to decisively intervene on the design of the administrative model, to exploit artificial intelligence technologies in the design and conception of services, and to completely rethink the governance process. Digital transformation is an ongoing process that requires frequent adjustments of its processes, services

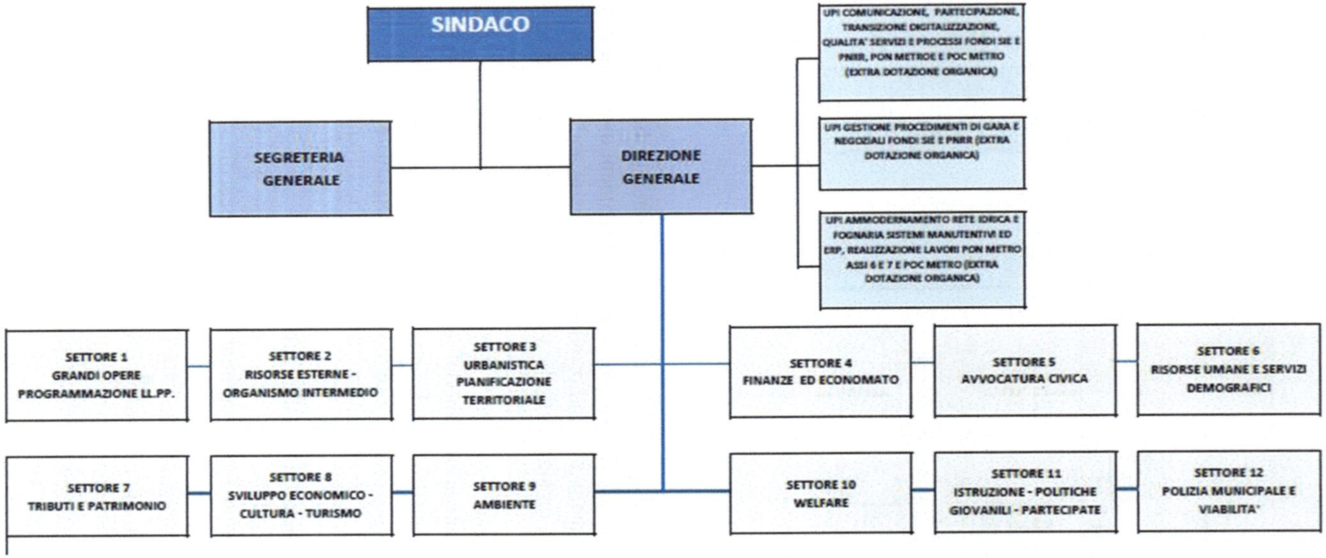

Fig. 3 Organisational structure chart of the city of Reggio Calabria

and products to external requirements, and experts are rarely able to point out what a digitally transformed public administration might look like because digital transformation is considered a process without an end status, unlike e-government projects [9]. This can be done by envisaging a transition macrostructure that assigns to a dedicated innovation directorate (digital, data science, artificial intelligence) the mission and the related digital transformation programs, making the necessary changes to the organisation of the entity and highlighting the weight of the policy choice and investment to be made. Within this intermediate macrostructure, it is therefore useful to envisage *in line with* the Director General a directorate, headed by a digital transition manager (RTD), dedicated to the overall coordination of the digital transformation that would take care of the actions of all sectors, centralising functions of planning, design and data collection and of cooperation with the national level to organise and graft the new solutions and prepare the digital model. This directorate would be in charge of managing the transition and preparing the model change. An example of organisational architecture is schematised in Fig. 4.

The office of the Head of Digital Transformation should have adequate expertise on public organisation reengineering processes and useful resources to be able to design and implement the transformation process, with particular regard to the creation of the data infrastructure and IT networks, the structuring of the municipal digital service platforms, the definition of the data *governance* process and the opening of data to the outside world, also to stimulate the active participation of citizens in the governance process in a subsidiary logic of *shared administration* [48]. The units of the Digital Transformation Directorate could be divided and structured according to the objectives of the process: competence building and general training, development of the digital transition programme, special projects (infrastructure, platforms and data quality), digitisation of existing procedures and creation of digital services (transition phase and digital transformation phase). Digital transformation is a comprehensive organisational approach and focuses on a holistic process to change products and culture, going beyond digitisation and digitisation, i.e. including the organisation as a whole and changing the bureaucratic and organisational culture [9]. The use of artificial intelligence and data science implies a paradigmatic or third-level

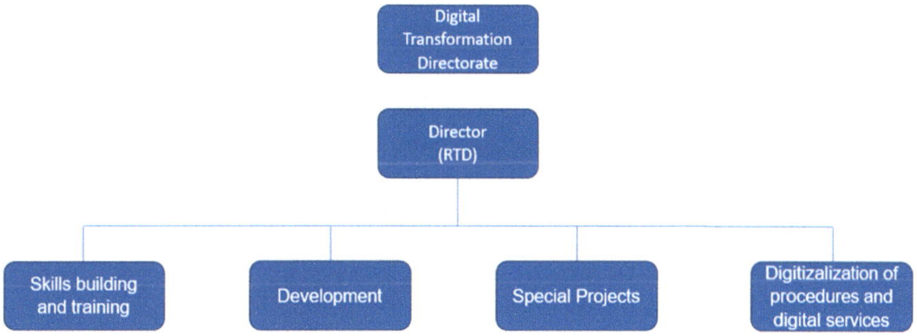

Fig. 4 Organisational chart of the Digital Transformation Directorate

change in the public policy cycle, bringing about a change in the principles, design, timing, tools of the cognitive scheme of the process and, consequently, its ability to generate results [1]. The entire process cannot follow a technological direction and change management must prioritise the administrative and organisational dimension that is the enabling context of public administration innovation processes.

The first step that local authorities have taken in the digitalisation process is a cultural step and at the same time a regulatory adjustment (CAD), namely the process of reorganisation, simplification and digitalisation of administrative procedures and sub-processes. This process starts with the definition of a manual of all the administrative procedures of the entity, useful for internal activity and the functions and services of competence. The Manual of Procedures must, however, once the first phase of mere digitalisation has passed, translate into the dematerialisation platform for the management and monitoring of all procedural tasks, which represents the single point of access for any activity, request, measure, or communication from and to the body, which thus become digital administrative procedures. At the same time, the authority must invest in changing the model by implementing the infrastructure that collects the data collected from heterogeneous sources. The process of acquiring the databases and creating the datasets used in the algorithmic sequence corresponds in some respects to a true administrative inquiry in that, the unstructured data stored in the *data lakes* must first be structured and analysed in order to ensure quality and relevance and then need Advanced Analytics algorithms to be able to generate predictive analysis and data discovery operations. The data infrastructure of an organisation is therefore the heart and brain of the system and the activation of the macro services of each organisational structure starts from the processing and transformation on the data lake through a Data Management process to transform the data into information that the IT ecosystem will use.

5 The Digitization of the Urban Planning Sector of the Municipality of Reggio Calabria

In 2016, the Urban Planning sector[2] of the Municipality of Reggio Calabria defined an enterprise technology platform through BPM software to digitise and manage the business processes of the Urban Planning sector, defining the digital administrative procedures of the Commencement of Ascertained Works Notice (CILA) of the End of Works Procedure and the Certified Declaration of Agility (Segnalazione Certificata di Agibilità—SCA) [49, 50, 51]. Starting from the Manual of Procedures in use at the authority since 2000 and the result of an initial phase of public organisation reengineering that aimed at outsourcing some services and administrative

[2] Town Planning and Smart City Councillor Agata Quattrone together with project manager Filippo Derrigo defined the digitalisation project of the three procedures as part of a broader digital reorganisation programme of the authority.

functions through the computerisation of administrative procedures [24], it rationalised the work of the offices, transforming both the traditional paper intake of practices from the outside through the counter, characterised by very low digitalisation, and the document management process that was previously non-automated and highly inefficient. Therefore, it was not simply a matter of digitising the form of the various applications, but of totally dematerialising the procedure by exploiting all the standard technologies (SPID, PAGOPA, Digital Signature/Stamp, Dematerialised Revenue Stamp) to eliminate any paper production from submission, to intermediate steps, to the final measure, allowing constant monitoring both for the user and the sector manager. Through BPM (Business Process Management) the authority has automated the submission/creation, assignment, monitoring, management and conclusion of each request and its internal flows, achieving an improvement in process efficiency (performance measurement through KPIs), in the productivity of the Urban Planning Sector (reduction of errors and human work), in the time of the procedure and in security (of public data and users' personal data) and administrative transparency (continuous monitoring). It is envisaged that the integrity and security of documents and proceedings will be ensured by *notarisation* on a private Blockchain to achieve an immutable digital register. The use of Blockchain results in an orientation towards the administrative transparency of the entire process, also allowing rapid access by users to information and the elimination of any intermediation through the automation of many operations, operated by *smart contracts* within the software. The entire process has been defined in compliance with Law 241/1990 (on administrative procedure and the right of access) and with Presidential Decree 380/2001 (Testo Unico sull'Edilizia), which provides for national standardisation of building forms. The platform allows for a fast and rational management of applications, guaranteeing at all times the transparency and traceability of the procedures submitted. In fact, each user has the possibility of following and checking the status of the dossiers under his or her responsibility step by step, verifying the stages of investigation and any requests for documentation or additions. The digitisation of the 3 procedures has already had a significant impact on the organisation of the sector, which this work has measured in terms of productivity and rationalisation of human resources. According to the analysis carried out by the present work following an audit of the offices' production in the first half of 2024, the organisational innovation already introduced in the authority with the first phase of digitisation of the three building procedures made it possible to partially compensate for the conspicuous decrease in the personnel available to the sector ($-$ 61.5% compared to 2008) generated by the restrictions imposed by a long debt recovery plan that the authority had to endure and that prevented the coverage of turnover. In 2023, the sector had 37 staff members and accumulated delays in carrying out its functions. The numbers that can be seen from the 2008–2023 comparison (74 units vs. 27) are dramatic, and some offices were not even able to carry out their ordinary activities. It was calculated that the sector, in light of the innovation process described above, could recover an adequate standard of service with an increase of 15 staff members to 42 (a reduction of 43.3% compared to 2008), achieving the same results as in 2008 when 70 staff members were in service. By means of a quantitative analysis of the administrative procedures falling within the competence of each unit, this work also proceeded to a simulation of the tendential staffing needs associated with a next step aiming massively at digital transformation. With a complete digitisation of the sector combined with the use of AI technologies, the management of the sector would be ensured with just 30 units and thus a 59.5% reduction in staff compared to 2008. The following table describes the placement of personnel in the various units in the three hypotheses described:

ARTICOLAZIONE UFFICI E DISTRIBUZIONE PERSONALE

Modello Tradizionale Vs in Transizione Digitale

Personale in servizio 2008-2024-futuro modello digitale

URBANISTICA PIANIFICAZIONE TERRITORIALE	Area degli istruttori	Area degli operatori	Area degli operatori esperti	Area dei funzionari e dell'EQ	Dirigenti	Totale anno2008	Totale anno 2024	Delta Modello in Transizione	Totale Modello Digitale
ATTUAZIONE PAU-P.d.C.-P.d.C. CONVENZIONATO [ufficio]	5	1		1		7	3	2	1
CERTIFICAZIONI URBANISTICHE ED EDILIZIE (INCENTIVI FISCALI)	3					4	1	1	1
CILA - SCIA - SCA - SUAP [ufficio]	3	1				4	2	1	2
CONDONO EDILIZIO [servizio]							1		1
CONDONO EDILIZIO [ufficio]	5			1		8	3	2	1
CONTROLLO SUL TERRITORIO [ufficio]	1	1				2	2	3	1
ESECUZIONE ORDINANZE [ufficio]	2					2	1	1	1
FRONT OFFICE SPORTELLO UNICO EDILIZIA [ufficio]	4		1	1		6	2	1	2
GESTIONE AMMINISTRATIVA ATTI CONDONO [ufficio]	1	1		1		3	1		1
GESTIONE ATTI E RAPPORTI CON LE AUTORITA' [ufficio]	3	2	1	1		7	4		4
GESTIONE E ATTUAZIONE PIANO STRUTTURALE [servizio]	4	2		1		7	2		1
macroarea PIANIFICAZIONE TERRITORIALE	1	1		1		3	2		1
SEGRETERIA AMMINISTRATIVA [ufficio]	3	2	1	1		7	4		5
UFFICIO CONTABILE [ufficio]	2	2		1		5	1	1	1
UFFICIO DI PIANO / SIT / URBAN CENTER [ufficio]	2	2		1		5	2	1	1
URBANISTICA E EDILIZIA PRIVATA [EQ]				2		2	4		2
VALORIZZAZIONE DEMANIO MARITTIMO [servizio]			1	1		2	4	1	2
VIGILANZA DEMANIALE [ufficio]				1		1	2	1	1
VIGILANZA E PUBBLICA INCOLUMITA' [servizio]	6	2		1		1	1		1
	11	8	4	9	1	74	42	15	30

Fonte: Elaborazione P4C

Let us reflect on three figures that are significant for reflection. The number of staff in service in 2008 before partial digitisation (74), in 2024 (42) and the estimated number of staff for a fully digital and AI-functional model (30). The two perspectives (transition model and fully digital model) are based on a number of workload simulations for the individual services undergoing technological upgrading that are the subject of this paper and verified by a field test.[3] The indication that emerged is that by recovering 15 out of the 44 units lost since 2008, the urban planning sector achieves similar performance, while through a massive use of new technologies it could further reduce its human resources allocation to 30. This would require a complete rethinking of the architecture of the authority designed on the basis of a culture of data management, to build data-driven organisation, programs and planning. The extension of the work set up in 2016 to all the procedures and offices of the Urban Planning sector and then of the entire municipality would be the general application of the pilot process and the related change and training necessary for the new model. In Fig. 5, the transition process is schematised with a simplification of the digital and AI technologies that are most likely to have an impact on each existing organisational unit (without making a change in the set-up).

[3] The simulation took into account the production of the Urban Planning sector, the workloads of each unit, productivity after digitisation of administrative procedures, and the possible impact of certain artificial intelligence techniques in the individual services of the entire municipal sector.

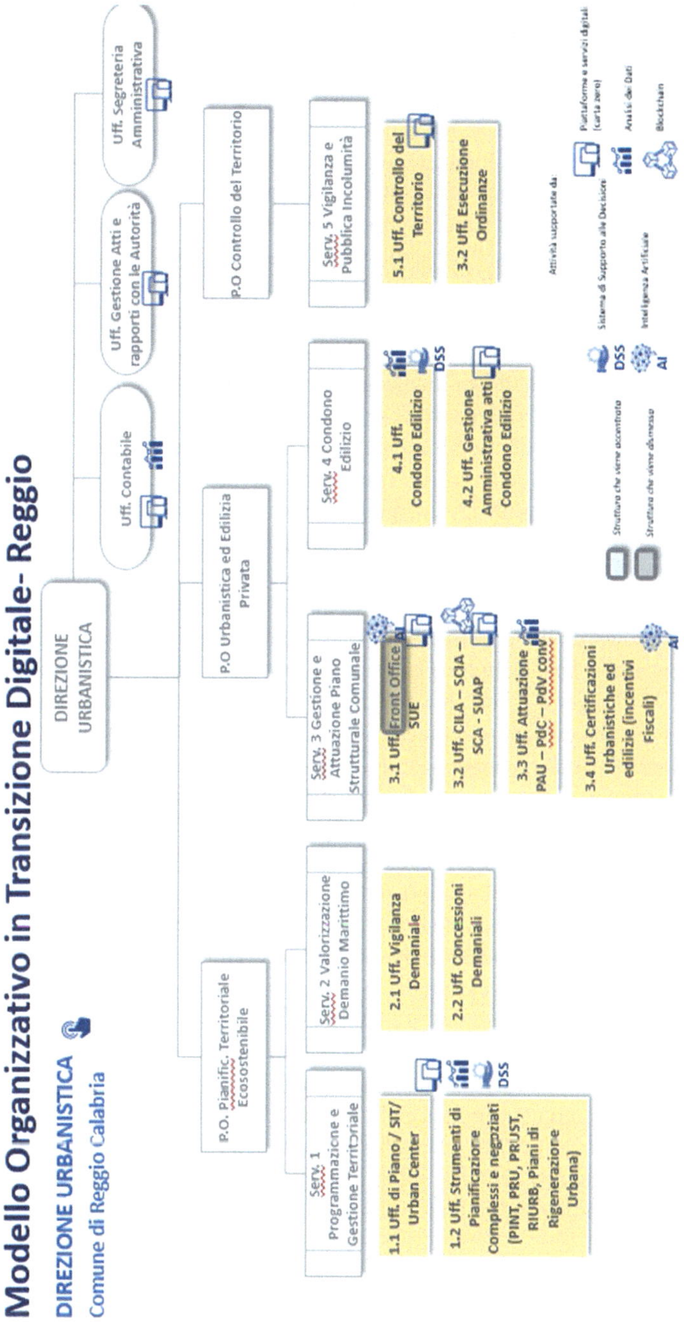

Fig. 5 Organizational Digital Transition chart of the city of Reggio Calabria

Artificial Intelligence technologies, DSS and data analysis systems, digital platforms, the use of drones to control the territory, Digital Twin models[4] if applied to the sector would allow a radical rethinking process of the sector's organisational structure with further advantages in terms of efficiency, productivity and quality of services. The analysis of the spin-offs of the innovation process revealed some critical issues. Part of the bureaucracy initially resisted the introduction of the new procedures, finding support in some stakeholders. The reasons of the bureaucracy were heterogeneous and in some cases due to the increased level of transparency and control of the activity with the consequent reduction of bureaucratic discretion in the management of priorities. The perplexities of some professional associations were, on the other hand, due to the usual fears about the switching costs of change, in terms of time and effort to be devoted to the transition to the new model and system. Currently, other administrative procedures of absolute importance (Building Permit, Certified Declaration of Commencement of Activity and Conditional Assevered Work Commencement Notice) are hosted on a regional platform (CalabriaSue), which is not interoperable with the Municipality of Reggio Calabria. This strabismus and lack of cooperation between territorial authorities in the flat-planning of services, highlights a defect in the multi-level organisation of the digitisation process and the shortcomings in terms of public sector governance, as well as the relevance in innovation processes of the aspect of organisational reengineering.

If we frame the digitisation of procedures as the first step in the transition process, the second step, towards the construction of a digital model, should start after the training of new tools and technologies, scaling up innovations and following a different path. No longer the search for a digital adaptation of the existing model and procedures, but the design of digital native e-services, followed by the definition of new digital administrative processes, and then designing on these processes an organisational structure that enhances the potential of the new technologies and is functional, not to a rigid and abstract scheme of specialisation, but is sensitive to the political-administrative objectives of each administration. Such a process would be functional to an administration of results and the production of public value and would break the rigid schematism of organising offices by silos, allowing the organisation to be dynamically adapted to the administration's programme. In the representation (Fig. 6) we compare the sequences that the two described processes presuppose. That of mere digitisation and that of digital transformation.

Figure 7 describes the design logic of a model that overcomes the rigid classification of offices, vertical by hierarchy and horizontal by competence, and enhances data and missions in a circular, immaterial, dynamic and data-driven PA logic.

Heterogeneous sources provide databases to the infrastructure that qualifies them into a data lake operated by artificial intelligence algorithms. The policy agenda is data-sensitive and technical, administrative and financial programming and planning

[4] The National Digital Twin (NDTp) programme was run by the Centre for Digital Built Britain, a partnership between the University of Cambridge and the Department for Business, Energy and Industrial Strategy of the UK Government Treasury. It represented a key step in the digital transformation of the infrastructure and construction sectors to promote change and embrace connected digital twins within organisations.

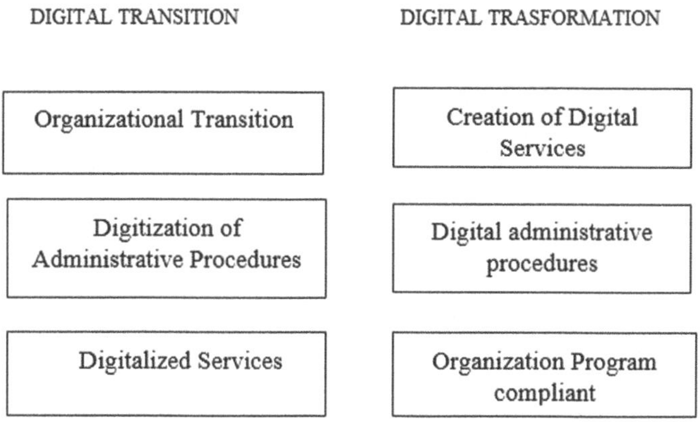

Fig. 6 Diagram representing Digital Transition vs. Digital Transformation

becomes precision, the operational management of the bureaucracy is dynamic and non-mechanical and translates adaptive and proactive policies, the result is precision i.e. less constrained by the traditional implementation deficit of theorised policies without databases.

6 Discussion: The Algorithm Between Transition and Digital Transformation

In order to optimise the application of AI techniques in the field, it is necessary, to invest in the data infrastructure, to make the data become process and immediate its integration in the service of the political-administrative programme, policy planning and its operational implementation and management. Mere digitisation and transformation by mere technological addition leave the culture and model of government (doctrinal cognitive scheme, hierarchical decision-making, rigid organisation design) substantially untouched and thus also maintain the form and material image of the traditional administration and downgrade innovation to a mere computerisation of bureaucratic administration processes. Instead, the digital transformation of the administration necessarily entails an organisational transformation of offices to adapt to a new decision-making process. A profound methodological rethinking, capable of delineating a digital and immaterial public sector, which outlines new services and new processes, defines new digital administrative procedures and not mere transpositions from the analogue scheme and structures an organisation that is not rigid but on the contrary dynamic. Sectoral innovations, although scalable after the pilot project, may not be disseminated or may not generate the expected public value due to various factors and barriers (political, cultural, bureaucratic, transition costs). The main barrier in the case of technological innovation is the maintenance of an organisational model designed for analogue administration that is functional

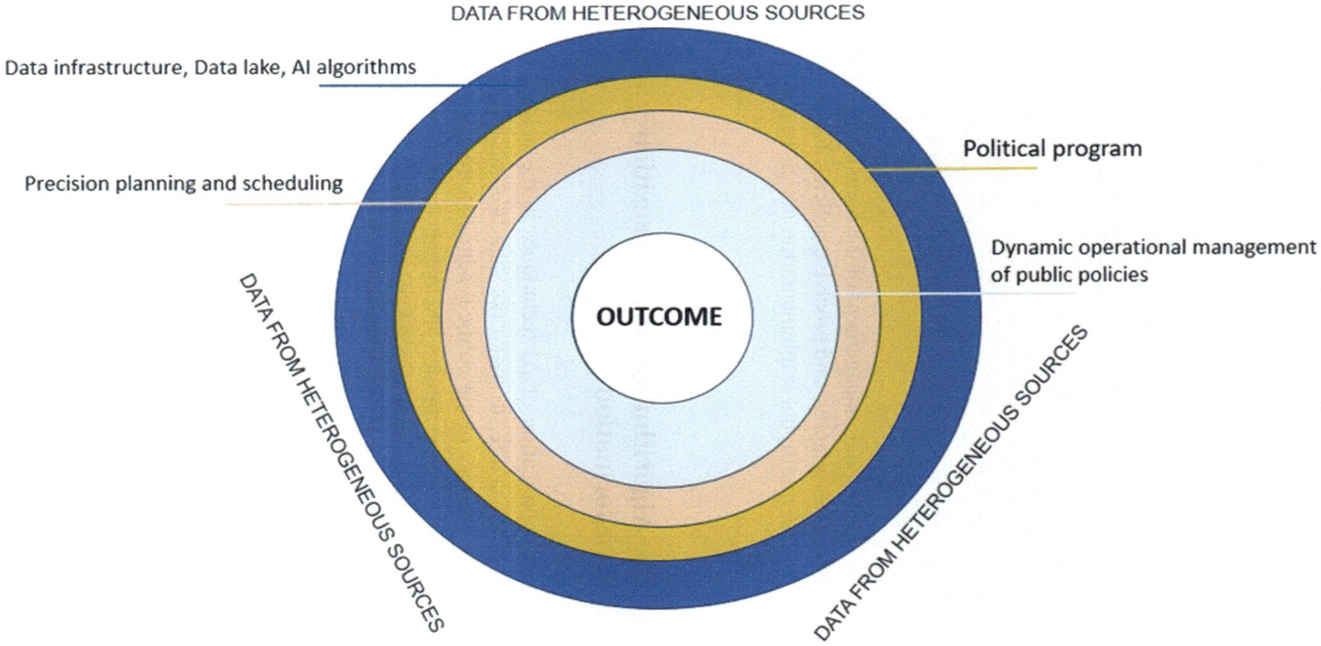

Fig. 7 Circular diagram illustrating data driven model for Public Administration

to different and exclusively human cognitive and decision-making paths. Innovation and organisational rethinking stand and fall together. Trivially, a new model based on the knowledge of data and the capacity of algorithms could not express itself without a framework of public action that enhances them. However, innovation applied to individual segments of public administration can generate even appreciable localised productivity benefits, as in the analysed case of the Urban Planning sector. For an algorithmic precision capability, after a transition phase useful to understand artificial intelligence techniques, a new administrative paradigm is needed to rethink modalities and tools to devise new digital services that, together with the data infrastructure, allow the Digital Transformation Directorate to plan a rethinking of the organisation of the authority towards an immaterial declination. Political and bureaucratic commitment must not be satisfied with partial results (as in the analysed case of the Urban Planning sector) that can serve to compensate for the lack of human resources. If it is even useful to train for innovation with a transition model (stages, actions, training, skills) that gives effect to the guiding principles of the Digital Administration (in Italy C.A.D. Legislative Decree 82/2008) guaranteeing records management, it is a digital architecture that conforms to and is functional to AI technologies that allows processing, predicting, supporting citizens, bureaucracy and political decision-makers, making evident what a traditional reading does not grasp. The organisation of offices and services is the circulatory apparatus of the system and must be designed on the basis of the algorithmic component as an essential element of change. It is artificial intelligence that gives the public sector a new algorithmic public capacity for *precision*, paradigmatically different in heuristic and implementation capacity, design and administrative culture. Without this capability, the public sector is destined to be marginalised and to undergo processes of substitution that will make it irrelevant. The lack of reengineering, rather than a deficit in organisational implementation, probably conceals the removal of the irruption of the machine (algorithm) alongside the human in the processes of analysis, conception, prediction and decision formation. Removing the Algorithm as a strategic element of the administrative process and thus not integrating it into a functional organisational framework, however, leads to the widening of the gap and the temporal dystonia between the public sector and society. The artificial intelligence algorithm must be defined (made transparent by design) and culturally introjected into the administrative process. The definition of its role will connote the legal and organisational image of the public sector. A mere tool, a referent of the decision-maker, a trained virtual party, an advisor (good or bad), a big brother? Much will depend on the public sector's ability to equip itself to comply with technology and artificial intelligence and not fall further behind the growing gap with the private sector and its repercussions with respect to Public Function as well as Public Decision. The pervasiveness of technology, however, requires from legal analysis a processing pace consistent with the speed imposed by technology. Only by rethinking the model of policy construction, delivery, and implementation in the light of AI and digital technologies is it still possible to imagine a Public Administration and thus a Democracy that is *precision* and thus effective, subsidiary, and shared.

Acknowledgements We would like to thank the former Councillor for Urban Planning and Smart City of the Municipality of Reggio Calabria, Eng. Agata Quattrone, and Dr. Filippo Derrigo (project manager of the sector digitisation project) with whom we had a fruitful discussion in relation to their work on the digitisation of services in the Urban Planning sector and with respect to the issues of digital transformation of the public sector.

References

1. Naccari Carlizzi D, Quattrone A (2022) Artificial intelligence and data governance for precision ePolicy cycle. In: Marino D, Melchiorre M (eds) Artificial intelligence and economics: the key to the future. Springer, Cham. ISSN 2367-3370, E-ISSN 2367-3389
2. Wirtz BW et al (2020) The dark sides of artificial intelligence: an integrated AI governance framework for public administration. Int J Public Adm 43(9):818–829. https://doi.org/10.1080/01900692.2020.1749851
3. Medaglia R et al (2021) Artificial intelligence in government: taking stock and moving forward. Sage J 41(1). https://doi.org/10.1177/08944393211034087
4. Avanzini G (2022) Intelligenza artificiale, machine learning e istruttoria procedimentale: vantaggi, limiti ed esigenze di una specifica data governance. In: Pajni A et al (eds) Intelligenza artificiale e diritto: una rivoluzione? Vol. 2: Amministrazione, responsabilità, giurisdizione. Il Mulino
5. European Commission (2021) Coordinated plan on artificial intelligence. COM (2021)205, 21 Apr 2021
6. Zuiderwijk A et al (2021) Implications of the use of artificial intelligence in public governance: a systematic literature review and a research agenda. Gov Inf Q 38(3):101577. https://doi.org/10.1016/j.giq.2021.101577
7. AGID (2024) Piano triennale (2024–2026) per l'Informatica nella Pubblica Amministrazione. del Governo italiano
8. Naccari Carlizzi D, Quattrone A (2021) Metropolitan cities and digital agenda: strategy and monitoring methodology. In: Bevilacqua C, Calabrò F, Della Spina L (eds) New metropolitan perspectives. NMP 2020. Smart innovation, systems and technologies, vol 178. Springer, Cham. https://doi.org/10.1007/978-3-030-48279-4_97
9. Mergel I et al (2019) Defining digital transformation: results from expert interviews. Gov Inf Q 36(4):101385. ISSN 0740-624X. https://doi.org/10.1016/j.giq.2019.06.002. https://www.sciencedirect.com/science/article/pii/S0740624X18304131
10. Aoki N (2020) An experimental study of public trust in AI chatbots in the public sector. Gov Inf Q 37(4):101490
11. Kuziemski M, Misuraca G (2020) AI governance in the public sector: three tales from the frontiers of automated decision-making in democratic settings. Telecommun Policy 44(6):101976. ISSN 0308-5961. https://doi.org/10.1016/j.telpol.2020.101976
12. Samoili S et al (2020) AI watch. Defining artificial intelligence. Towards an operational definition and taxonomy of artificial intelligence. (JRC118163) Technical report. Joint Research Centre (Seville site). https://publications.jrc.ec.europa.eu/repository/handle/111111111/59452
13. Matheus R et al (2020) Data science empowering the public: data-driven dashboards for transparent and accountable decision-making in smart cities. Gov Inf Q 37(3). ISSN 0740-624X. https://doi.org/10.1016/j.giq.2018.01.006
14. Naccari Carlizzi D et al (2023) Decision making and e-Democracy: Archimede, a tool to support the data-driven process. In: Marino D, Monaca MA (eds) Innovations and economic and social changes due to artificial intelligence: the state of the art. Studies in systems, decision and control, vol 222

15. van Noordt C, Misuraca G (2022) Artificial intelligence for the public sector: results of landscaping the use of AI in government across the European Union. Gov Inf Q 39(3):101714. ISSN 0740-624X. https://doi.org/10.1016/j.giq.2022.101714. https://www.sciencedirect.com/science/article/pii/S0740624X22000478
16. Casonato C, Marchetti B (2021) Prime osservazioni sulla proposta di regolamentodell'Umione europea in materia di intelligenza artificiale. Bio-Law J-Riv BioDir (3)
17. Chiti E et al (2022) L'impiego di sistemi di intelligenza artificiale nelle pubbliche amministrazioni italiane: prove generali. In: Pajni A et al (eds) Intelligenza artificiale e diritto: una rivoluzione? Vol. 2: Amministrazione, responsabili-tà, giurisdizione. Il Mulino
18. Galetta D-U, Corvalán JG (2019) Intelligenza Artificiale per una Pubblica Amministrazione 4.0? In: Potenzialità, rischi e sfide della rivoluzione tecnologica in atto. Federalismi, n. 3/2019. federalismi.it. ISSN 1826-3534, 6 Feb 2019
19. Naccari Carlizzi D, Quattrone A (2022) The path to algorithmic public decision and the necessary contamination between law and information technology. In: Marino D, Melchiorre M (eds) Artificial intelligence and economics: the key to the future. Springer, Cham. ISSN 2367-3370, E-ISSN 2367-3389
20. Ferrari M (2022) Profili giuridici della predizione algoritmica. Guida Editori. ISBN: 979-12-5563-009-8
21. Fatima S et al (2020) National strategic artificial intelligence plans: a multi-dimensional analysis. Econ Anal Policy 67(C):178–194
22. Deidda Gagliardo E et al (2014) Public value measurement in cultural services: a case study. In: Guthrie J, Marcon G, Russo S, Farneti F (eds) Public value management, measurement and reporting. Third volume of the new Emerald series studies in public and non-profit governance. Emerald Books
23. Van Noordt C, Tangi L (2023) The dynamics of AI capability and its influence on public value creation of AI within public administration. Gov Inf Q 40(4):101860. ISSN 0740-624X. https://doi.org/10.1016/j.giq.2023.101860. https://www.sciencedirect.com/science/article/pii/S0740624X23000606
24. Naccari Carlizzi D (2002) Innovazione e sviluppo nei governi locali: il caso di Reggio Calabria. In: Manuali Anciform-Collana editorial. ANCI-Editrice C.E.L. ISBN: 88-7951-335-4
25. Crosetti A, Fracchia F (2002) Procedimento amministrativo e partecipazione. Problemi, prospettive ed esperienze. Giuffrè Editore. ISBN: 88-14-09721-6
26. Di Viggiano PL (2010) l'Amministrazione digitale. Progetti e tecnologie per l'e-government. Riv elettron dir econ manage 1:2010
27. Valle-Cruz D et al (2020) Assessing the public policy-cycle framework in the age of artificial intelligence: from agenda-setting to policy evaluation. Gov Inf Q 37. https://doi.org/10.1016/j.giq.2020.101509
28. Höchtl J et al (2016) Big data in the policy cycle: policy decision making in the digital era. J Organ Comput Electron Commer 26(1–2):147–169. https://doi.org/10.1080/10919392.2015.1125187
29. Rangone N (2022) Intelligenza artificiale e pubbliche amministrazioni: affrontare i numerosi rischi per trarne tutti i vantaggi. Bio-Law J Riv BioDir (2). https://doi.org/10.15168/2284-4503-2350
30. Vigoda-Gadot E, Mizrahi S (2024) The digital governance puzzle: towards integrative theory of humans, machines, and organizations in public management. Technol Soc 77:102530. ISSN 0160-791X. https://doi.org/10.1016/j.techsoc.2024.102530. https://www.sciencedirect.com/science/article/pii/S0160791X24000782
31. Lippi A (2022) Modelli di amministrazioni pubbliche. Il Mulino Manuali. ISBN: 978-88-15-29397-8
32. Lippi A, Morisi M (2005) Scienza dell'amministrazione. Il Mulino Manuali. ISBN: 88-15-09772-4
33. Cassese S (2022) Amministrare la Nazione, la crisi della burocrazia e i suoi rimedi. Mondadori. ISBN: 978-88-04-76634-6

34. Barzelay M, Gallego R (2006) From "new istitutionalism" to "istitutional processualism": advancing knowledge about public management policy change. Governance 19(4)
35. Di Mascio F, Natalini A (2018) Oltre il new public management, le riforme amministrative tra meccanismi e contesti. Carocci editore. COLLANA: Biblioteca di testi e studi (1165). ISBN: 9788843091461
36. Bobbio L et al (2017) Le politiche pubbliche, problemi, soluzioni, incertezze, conflitti. Mondadori Università, Milano. ISBN-10: 8861844669, ISBN-13: 978-8861844667
37. Cassese S (2024) Miseria e nobiltà, dialoghi sullo stato della Nazione. Solferino. ISBN: 978-88-282-1345-1
38. Merton RK (1952) Reader in bureaucracy. Free Press, Glencoe, IL
39. Battini S, Decarolis F (2019) L'amministrazione si difende. Riv trim dir pubbl 1
40. Abriani N, Shneider G (2021) Diritto delle imprese e intelligenza artificiale, dalla Fintech alla Corptech. Il Mulino Saggi. ISBN: 978-88-15-29428-9
41. Bonaccorsi A (2022) Dove abbiamo sbagliato, da dove ripartire. In: Cicione F (ed) Quattordici Lezioni sull'Innovazione. Rubettino editore
42. Torchia L (2023) Lo Stato digitale, una introduzione. Il Mulino Itinerari. ISBN: 978-88-15-29921-5
43. Van Dijck J et al (2018) The platform society. Public values in a connective world. Oxford University Press
44. European Commission (2020) Europe's digital decade: digital targets for 2030
45. Pajno A (2022) La costruzione dell'infosfera e le conseguenze sul diritto. In: Pajno A, Donati F, Perrucci A (eds) Intelligenza artificiale e diritto: una rivoluzione? Vol. 2: Amministrazione, responsabilità, giurisdizione. Il Mulino
46. Punzi A (2019) Judge in the machine. E se fossero le macchine a restituirci l'umanità del giudicare? In: Carleo A (ed) Decisione robotica. Bologna
47. Razzante R (2024) Il governo dell'intelligenza artificiale, gestione dei rischi e innovazione responsabile. Cacucci editore, Bari. ISBN: 979-12-5965-330-7
48. Cerulli-Irelli V (2023) L'amministrazione condivisa nel sistema del diritto amministrativo. In: Proceedings of the "L'Amministrazione condivisa" conference held by the University of Trento, Trento, 20 gennaio 2022. www.unitn.it/amministrazione-condivisa. Derrigo F (2022) La Dematerializzazione dei procedimenti amministrativi attraverso tecnologie Web 3.0 tesi di laurea in Scienza e Tecnologia dell'Informazione. Unipegaso
49. Marino D, Cananzi D, Aragona F (2024) Artificial intelligence and ethics challenges of technologies. In: Ethics and artificial intelligence. Studies in systems, decision and control, vol 519. Springer, Cham. https://doi.org/10.1007/978-3-031-50902-5_3
50. Marino D, Trapasso R (2009) The new approach to regional economics dynamics: path dependence and spatial self-reinforcing mechanisms. In: Fratesi U, Senn L (eds) Growth and innovation of competitive regions. Advances in spatial science. Springer, Berlin, Heidelberg. https://doi.org/10.1007/978-3-540-70924-4_15
51. Marino D, Trapasso R (2020) Servitization and territorial self reinforcing mechanisms: a new approach to regional competitiveness. Investig Reg—J Reg Res 48:85–93. https://doi.org/10.38191/iirr-jorr.20.022

Artificial Intelligence: Crucial Area of Geopolitical and Power Competition Between the West and Emerging Countries

Domenico Marino, Pietro Stilo, and Francesco Serra

Abstract This paper explores the critical role of artificial intelligence (AI) as a strategic frontier in the geopolitical and economic competition between the West, led by the United States, and emerging global powers, particularly China. It situates the current AI rivalry within a broader historical and economic context marked by shifting paradigms in globalisation, technological innovation, and power dynamics. The discussion highlights how AI development is reshaping global value chains, fostering new economic and social structures, and intensifying the dualism between the United States and China. Emerging economies, including the BRICS and African nations, are identified as pivotal players in the evolving multipolar world order. This study also examines the implications of AI for national security, economic strategies, and global governance, emphasising the need for multilateral cooperation to address the risks and opportunities of this transformative technology. By analysing recent developments and future projections, the paper contributes to understanding AI's role in shaping the 21st century's geopolitical landscape.

Keywords Global value chain · Globalisation · Emerging economies

1 Global Overview

The current phase of human history is particularly complex. We are going through a new paradigm shift, where many uncertainties, many instabilities have eroded situations that seemed cemented by time. Paraphrasing in reverse the famous phrase of the American political scientist Francis Fukuyama, we could say that we are at the return of history, which has probably never ended, in fact contradicting the theses contained in his prophecy, with which he sustained precisely the 'end of history'. With this statement, he was referring to the fact that, after the collapse of the Berlin Wall and the consequent end of the Soviet bloc and thus the Cold War,

D. Marino (✉) · P. Stilo · F. Serra
Mediterranean University of Reggio Calabria, Reggio Calabria, Italy
e-mail: dmarino@unirc.it

D. Marino and M. Monaca (eds.), *Generative Artificial Intelligence and Fifth Industrial Revolution*, Lecture Notes in Networks and Systems 880,
https://doi.org/10.1007/978-3-031-73880-7_7

democracy and Western-style economic liberalism would be the dominant political-economic ideologies throughout the world, which would gradually pervade the whole of humanity.

After a phase of 'interruption' in world history, about thirty years, dominated precisely by US-driven economic and ideological-cultural globalisation, during which Fukuyama's prophecy really seemed to come true, the international situation seems to have changed again, the world is once again pervaded by tensions, regional conflicts, geopolitical and economic challenges from new global players challenging the stars and stripes leadership, and so we can say that this prediction, made at a certainly positive historical moment for the US, was probably too optimistic, not to say imbued with too much 'western centrism'. It is clear that the American political scientist in making this statement had not taken into consideration the cycles of history and above all the fact that numerous factors of global geopolitical and economic dynamics cannot be analysed by considering events, however important they may be, dictated by a single event. In order to read what is happening on the planet and thus be able to give a clear and objective reading of it, long-term, multilevel and multicultural keys are needed. The proof of this statement is the evidence of how the world around us is changing again, a world that is increasingly disordered, competitive, complex and unstable, a polycentric world. We are going through a phase in history, where history is increasingly present and incisive, where numerous phenomena are bringing about changes even in the very perception of change. The consequences of globalisation, such as: galloping urbanisation, the economic-financial crisis of 2007–08, the ecological transition, the digital transition, the Arab Springs, regional geopolitical changes, deglobalisation, disruptions in global value chains, competitions between the hegemon and its challengers, first and foremost the increasingly strong challenge between the US and China (in particular for technological superiority, especially for artificial intelligence), the COVID-19 epidemic, the Russian-Ukrainian war, the resurgence of the Israeli-Palestinian conflict, the return of the BRICS (Brazil, Russia, India, China and South Africa) enlarged to include new members, the demographic growth of Asia and Africa, the tensions in the South China Sea (Taiwan and North Korea), the wave of coups in Africa, where the race to hoard resources is very disruptive, so we can see how history has returned (if it was ever over) and has once again traced itself, outlining an increasingly messy and complex world.

2 Globalisation in the Present Era

However, globalisation at the current stage seems to be at a standstill, probably due to some changes taking place, some attempts to bring back production that had been spread over several countries in several continents. As is well known, globalisation is a very complex phenomenon that consists of economic, social and cultural integration between various areas of the world, stimulated and driven by technological innovations that act as a driver. From an economic point of view, it is mainly a process of networking between companies located in different countries around the world and also the foreign projection of companies that move all or part of their domestic production in search of new markets for their products and services or to find cheap labour, raw materials and energy and save on transport costs, for example, to generate economies of scale and thus be more competitive. Globalisation has as its main actors not only multinational companies, but also and above all states and international economic-financial institutions, and it reverberates in the fall of barriers to international trade, in the increase in the exchange of goods and services between states, and in the expansion of international investments, both direct and indirect, and in the reduction of all those cultural and psychological distances that separate states and peoples. Drivers of globalisation, as mentioned above, are the technological innovations that have accelerated its effects and scope, but also the dismantling (after the Bretton Woods Agreements of 1944) of restrictions on international trade with the creation in 1947 of the GATT (General Agreement on Tariffs and Trade), which later became the WTO (World Trade Organisation) in 1995.

The impact of globalisation on the contemporary world has had and still has effects as few other events have had in recent decades. Its acceleration has recent origins, although the original causes have roots going back a long way in time and date back to man's need to exchange goods and compare notes, to make contact and communicate. All this gave rise in the past to the journeys of the great merchants in search of new goods and business, think of the epic of Marco Polo's voyages of a million, or the voyages of the English, Dutch, Portuguese and Spanish shipping companies to distant realities in search of spices and other merchandise, these are just a few examples, distant and evocative but real and concrete of what has happened over time. Technological innovations have played a fundamental role in enhancing the development of globalisation, which goes hand in hand with industrial revolutions, one thinks of the birth of steam engines that changed the production system with the so-called first industrial revolution, which in turn also changed the housing distribution of people who, out of a need for work, began to leave the countryside and head for the cities. In this context of displacements, innovations in the field of transport and communication are inserted, one thinks of the role of the train in this context, in fact it redesigned the way of travelling, transporting goods and conveying postal materials and united cities and the countryside, thus contributing to the formation of the national identities of countries such as Italy or the United States of America just to give a few examples [5].

Other determined innovations have been: the telegraph, radio, telephone, fax, television, satellite communications and the internet, some of which first originated in the military and were later transposed to the civil sphere. In recent times very impactful are: information technology, artificial intelligence, the wide diffusion of smartphones, tablets and PCs connected to the internet and the use of social networks that have increased the possibility of interacting as never before in the history of mankind, think for example of the replacement of letters with e-mails what impact this has had in terms of speeding up written communication, both for personal and work relations.

It is a phenomenon of epochal proportions, bringing together the most diverse subjects, reducing their distance both geographically and communicatively as well as culturally and psychologically. It has reduced cultural, institutional, political, economic and regulatory gaps between states. For its detractors, globalisation has instead flattened local identities by levelling the planet and exploited one part of the world for the benefit of the other [4, 8].[1]

According to geopolitical scholars, globalisation is primarily the external projection of the power of an empire (formal or de facto), the Roman Empire, then the British Empire, and lastly the unipolar United States, which de facto controls the seas and oceans through so-called bottlenecks (isthmuses, canals, straits, through which some 90% of global trade passes by ship) and thus dictates the rules of international economic exchange and hence globalisation [7]. In recent times, there has been a widespread belief that globalisation is in decline (in fact, people are starting to talk more and more about deglobalisation). Instead, globalisation is probably going through a phase of adjustment dictated by the changes in the world geopolitical balance that have taken place over the last decade.

3 Emerging Economies

In recent years, the role of emerging economies has become more and more incisive on the global scene. After a period of stalemate, the inclusion of new global alliances has revived the path of the BRICS, in fact in recent years they have regained a new activism thanks also to new memberships, in fact as many as 11 states have applied to join and six of them Ethiopia, Saudi Arabia, Iran, Egypt, the United Arab Emirates and Argentina have formalised their membership on 1 January 2024. The BRICS are the emerging economies, which an expert from the US investment bank Goldman Sachs, Jim O'Neil in November 2001 initially defined with the acronym BRIC (Brazil, Russia, India, China), which became many years later BRICS with the addition of South Africa. O'Neil [12],[2] in his analysis recommended these economies to his clients as good destinations to invest in, since in his opinion after a careful analysis of the macroeconomic and financial data, they would become competitive against those of the G7 (i.e. the industrialised countries), a thesis that was initially taken into little consideration, but which later found some acceptance. At the time of O'Neill's statements, these countries had no particular connection with each other, but a few years later, they began to come together and give shape to what is now a structured group that attempts to act as a counterbalance to the more advanced economies of the West also through the birth of economic-financial institutions such as the New Development Bank, in an attempt to assert itself against the balances

[1] Charles Hill, International Business, Economics and International Strategy: The Enterprise in Global Markets, Hoepli Ed. 2008.

[2] O'Neill, J., The Growth Map: Economic Opportunity in the BRICs and Beyond, New York: Penguin. 2011.

born after the Bretton Woods Agreements. With **Egypt and Ethiopia joining the BRICS and** many others aspiring to join including: Algeria, Nigeria, Morocco, etc. The role of the BRICS in Africa will become more and more incisive. The envisaged partnership between the African continent and the BRICS focuses on three specific areas: trade, foreign direct investment and development aid.

A potentially valuable cooperation that can foster economic growth, create jobs and accelerate the continent's inclusion in global value chains [3]. Results that can only come true by avoiding national appetites ***'Africa Must Unite'*, wrote Nkrumah, father of Pan-Africanism**.[3]

For example, between China and India there are rivalries for leadership in the Indo-Pacific, for some border areas, for the management of waterways such as the Brahmaputra River, one of the largest waterways in the world. China also geopolitically supports Pakistan and Nepal in an anti-Indian key, in turn India is supported by the US against China. Brazil, a vast country with large natural reserves, has among other things strong ethnic and cultural ties with the West, while it has religious ties with the USA, especially with American evangelicals.

The world has entered a new phase, in many ways different from the recent past. The Washington Consensus model of development has begun to lose its leadership. The US remains the most productive economy, despite the slow progress of other emerging economies, China above all. So what was considered the 'factory of the world' in the 1970s and 1980s, which worked on behalf of the West, producing the goods required by the markets of the developed countries at low cost, has at some point begun to take its destiny into its own hands, to look inward, to develop, to experience industrial development as it had done in the West a century earlier, to urbanise and produce for itself and consequently to compete with the West.

4 2024 the World on the Ballot

The year 2024 is a very important year, voting will take place in 62 countries around the world, a total population of about 4 billion people will be involved in this process, i.e. almost half of mankind. The countries involved are: United States, Great Britain, Iran, Taiwan, India, Indonesia, Mexico, South Korea, Pakistan, just to name a few. For presidential, legislative or administrative elections, **16 African countries, 11 Asian, 22 European, 9 American, 4 in Oceania will vote. In Europe, they will vote for the renewal of the Union's Parliament: a decisive test to understand in which direction Europe is heading**.[4]

If we are to reflect on the elections in America, Donald Trump's return to office may be premature, but in light of his record as the 45th President of the United States of America, the Tycoon of NY, during his old campaign, did not only criticise the approach and legacy that his antagonist, the outgoing Joe Biden, might leave

[3] Africa's challenges in the global world I Africa Magazine (africarivista.it).

[4] 2024, assault on democracy I ISPI (ispionline.co.uk).

behind, a heavy and contradictory legacy in his view for both America and the world, just as he did in the past when he criticised the courses of action of many American presidents from Barack Obama to the two Bushes, whom the former anti-system candidate considers to be a political caste like the Clintons, but even went so far as to criticise Ronald Reagan, who has always been considered a myth for the Republican electorate. In the past elections, behind his slogan 'America First' repeated almost like a mantra, Trump has always concealed his desire to change US foreign policy, for example by deprioritising NAFTA (a regional economic free trade agreement with Canada and Mexico) (his aversion was clearly directed at Mexico because of illegal immigration, drug trafficking and the cheap labour that attracts the relocation of US companies, which he instead tried to bring back home), just as his desire to regulate finance, which in his view from the Clinton presidency onwards has produced unprecedented inequalities, was notorious. Abroad, he has always advocated a different kind of engagement in the Middle East, perhaps shared with other actors; in fact, in the past legislature he eroded the already critical relations with Iran, through the ISA (Iran Sanctions Act) an extension of economic sanctions against Tehran (with widespread discontent among big companies such as Boeing). The reduction of American military intervention in the world could be brought back into the limelight once again, a reduction in commitment which, to tell the truth, was already beginning to make its effects felt under Obama, and even under Biden has seen a number of positions taken (e.g. withdrawal from Afghanistan), and in line with this thought is the redefinition of the role and economic contributions of the USA to NATO, towards which it has always asked its allies for greater economic commitment. Within this perimeter, however, are also relations with China, which have never been idyllic for Trump; in fact, he has implemented a policy of strong opposition to the Red Dragon, raised protectionist walls through tariffs and other trade policy instruments, with the idea of pushing US companies not to relocate, with Trump in the White House there is more and more talk of reshoring, nearshoring or friendshoring, i.e. of a return or at least a rapprochement of global value chains towards the US or like-minded countries.

5 Global Value Chains in the Age of Artificial Intelligence

Artificial Intelligence is already changing the world and the way many activities are managed, the speed with which production processes can be scaled up, but also the ability to increase the offer of innovative products and services, thanks to very flexible processes and a greater ability to connect with other digitised companies for a frequent exchange of data and information, and finally the ability for continuous improvement and learning (machine learning). Technology can thus become a long-term innovative process asset by interfacing with all the latest generation products such as robotic arms, cobots, AGVs, ATRs, and give rise to new ways of operational management.[5]

We are witnessing a paradigm shift, from a traditional factory model we are moving gradually but also quickly (especially in some sectors) to a digitised and technology-driven model [6, 9, 14].

From the Trump era onwards, it has been accentuated by the COVID-19 crisis and the disruptions due to causal incidents and yet to the Houti attacks in the Gulf of Beb al-Mandeb, which have in fact put global trade in crisis, highlighting all its weaknesses and criticalities. All this will change the hitherto known scenario, which is now understandable even to the uninitiated, but most probably without any particular repercussions, for several reasons, including the pragmatism that drives Donald Trump's political action. What is certain is that any space left vacant by the Trump administration will immediately be occupied by other global geopolitical players, for example by the so-called emerging economies of the planetary chessboard, which are very dynamic in this direction; together with them, the European Union itself could certainly play a greater role (in a potential scenario of this kind) if it were capable, however, of being more concrete and incisive in its external action and in its common objectives. In this key, we can note the line of 'multi-alignment' of the African continent, a fact to be taken into great consideration is **the entry of the AU (African Union) among the permanent members of the G20, a significant event in terms of unity, but also of a qualitative leap for the continent, this event will be able to** make its specific weight count more on the global chessboard.[6] An interesting reasoning in this sense, is set out in the recently published "The Potential of Africa to Capture Technology-Intensive Global Supply Chains" by UNCTAD (United Nations Conference on Trade and Development), in which the potential of the African continent to become a strategic and attractive hub for large international companies (but not only them) with regard to global value chains, supplies and in the future also distribution, is represented. According to the UNCTAD Report 2023,[7] Africa is home to raw materials and partly also semi-processed goods (the latter market could expand decisively) that are fundamental especially for chains related to new technologies, such as batteries for electric cars and the electric car sector in general, for mobile telephony, and in general in the high impact technology sector. An additional benefit would also be the geography that allows the continent to expand goods, semi-finished and finished products to every direction of the globe, with a potentially huge workforce if properly educated and trained. The United Nations estimates that by 2100 the world's population will be about 10.9 billion and the only growing continent will be Africa with about 3.9 billion people, in perspective this makes us well understand the importance of this reasoning. The African Free Trade

[5] WEF_A_Global_Rewiring_Global_Value_Chains_2022.pdf (weforum.org).

[6] Africa's challenges in the global world | Africa Magazine (africarivista.it).

[7] https://unctad.org/publication/economic-development-africa-report-2023#:~:text=UNCTAD%20calls%20for,-1&text=International%20organisations%20to%20support%20African,in%20technology%2Dintensive%20supply%20chains.&text=African%20countries%20to%20strengthen%20intellectual,to%20attract%20more%20foreign%20investment.

Area (AFCCFTA) is expected to increase intra-African trade by more than 50%, with a major spill-over effect on trade with the rest of the world, respectively 29% of exports and more than 7% of imports.[8]

6 The New Geopolitical and Economic Confrontation Between the US and China

At such a dramatically difficult time for the latest escalating international tensions, it is reasonable to consider of paramount importance the bitter and opposing Sino-US dualism, inevitably the needle of the scales and the gauge of the geopolitical and geostrategic temperature worldwide.

The vexata quaestio stems from the unthinkable gap between the incumbent geopolitical hegemon (the United States) and the now rising power (China), which has been challenging the global order in place since the end of World War II. The US-China rivalry originated in 1949 with Mao's revolution and the birth of the People's Republic of China of clear communist inspiration. The US in anti-communist China supported the nationalist government in exile in Taiwan of Chian Kai Shek, who fled after defeat in the civil war, the US aim was to counter the rise of communism in China and since then relations between the two countries have always been characterised by tensions, except for a brief pause during the early 1970s. Trade relations resumed in 1979, opening up a relationship that would give rise to trade worth billions of dollars; already in the first year after the agreement, trade reached 4 billion, today it is around 600 billion. China's economic rising has been dictated by many factors, including the decision to relocate the production of many Western companies in search of low labour costs and few regulatory rules in the 1970s, a large and growing population until 2023 when it was overtaken by India as the most populous country in the world, the ability to create the conditions to attract Western companies with tools such as SEZs, the entry into the WTO in December 2001, all of which increased production and therefore jobs and welfare and therefore domestic demand for goods and services, in fact in the last thirty years it has grown economically exponentially, just think that China's GDP today is about 19.4 trillion dollars second only to that of the USA which is 26.9. In 2012, the US, Europe and Japan unite in demanding explanations for China's restrictive policies on exports of rare metals, pivotal raw materials for semiconductor manufacturing, policies that violated free trade rules and pushed Western companies to locate in China to avoid such restrictions. 2012 was also the year of Xi Jinping's arrival as president of the People's Republic of China, who arrived with a policy focused on renewal and opening up to foreign markets. In 2014, after an attempted rapprochement between Xi and Obama, tensions rose between the two giants following the US accusations against five Chinese hackers of carrying out industrial espionage against American companies, triggering chain

[8] Africa's challenges in the global world | Africa Magazine (africarivista.it).

reactions between the two states. China immediately broke off cooperation on cyber-security TMS, but it was in 2018 with Donald Trump that this clash reached its climax. Despite an initial attempt at a trade dialogue, Trump and his Treasury Secretary start a trade war with China by imposing very stringent tariffs to which China in turn responds by applying tariffs on US goods. Trump's goal was to rebalance the trade imbalance with China as far as possible with a negative balance for America of USD 418 billion out of some 600 billion in total value.

Rivalry between the US and China is manifesting itself over a number of disputes that are essential to their state bases, notably with a trade war, technological conflict and the balkanisation of global supply chains.

As is now well known to most, the real challenge for technological supremacy today and tomorrow is the ability to understand the most complex and abstruse global phenomena. It is a story in which the security of states as large as continents intersects, at the risk of overlapping, with the business opportunities of companies and the hope for economic progress of hundreds of millions of individuals.

This being the case, there is no doubt that the design, use and development of artificial intelligence systems, hence the conception of technological dominance, may represent one of the instruments of geopolitical prevarication in today's global scenarios, already showing signs of its impact on both economic strength and future policies of a securitarian matrix.

Already since 2015, in the context of dutifully questionable economic espionage activities, with the complacency of *insiders* acting as '*insider threats*' or of shielded companies of the Republic in universities, in companies operating in the field of critical infrastructure, even in scientific and industrial research institutes, China was and is moving towards the goal of extricating itself from the American stranglehold in the field of the development of advanced integrated chips, even through the allocation of a huge sum of money (about 50 billion dollars) to ensure its self-sufficiency after it was restricted in semiconductor sales by the aforementioned adversaries.[9]

Just as it is above all on the terrain of artificial intelligence and the development of legacy chips that one glimpses the will to excel of the United States and China, the current and undisputed 'algocratic' superpowers, both contenders in a challenge as fiercely heated as never before in the history of our world.

In the pursuit of its goals, the AI is a particularly complex and difficult playing field, in the presence of which we witness a re-enactment of the Cold War, with the Western bloc headed by the United States and the NATO countries, on the other hand no longer the Soviet Union, but China and the BRICS Plus group, which try by all means to take the lead without any misunderstanding of the aims.

[9] From the case of the Chinese company ZTE, passing through HUAWEI up to 2023 with the case of Advanced Semiconductor Materials Lithography (ASML), China has not stopped economically attacking the most peculiar features of the sector in which the United States still holds the supremacy. For a clearer picture, read the Report of the same Company https://www.asml.com/en/investors/annual-report/2022; as for the subject of industrial espionage and the techniques used by the Chinese security systems, see A. TETI, China *intelligence Tecniche, strumenti e metodologie di spionaggio e controspionaggio della Repubblica Popolare Cinese*, Rubbettino, 2024, pp. 83 ff.

It is certainly one of the challenges for both Superpowers to project their authority and power in the Eurasian landmass to the sea-oriented democracies, and not to postpone it.

In this sense, the risk for the United States in particular is the possible crisis, harbouring the fierce certainty that it will no longer be able to 'defend the empire' in this respect.[10]

The dynamics taken from the famous volume 'War without Boundaries' should make us doubt whether we are in fact faced with possible 'weapons without boundaries',[11] provided that frontal confrontation even as part of a proper war strategy is not a real and possible priority at the moment [13].

Specifically, for the US, the real military guarantor, the measurement of one's own heartbeat is precisely aimed at maintaining primacy over technological developments tout-court.

Well, it has been announced, not least because of the trip by the American Undersecretary Antony Blinken, that an ad hoc round table will be set up to define the aforementioned application standards, i.e. try to define the rules for the use of artificial intelligence. At the moment, America on chips wants to contain Chinese progress, those of the highest level, while China, on the other hand, wants to strengthen its position in the area of legacy chips, i.e. those necessary for the production of common electronic tools, which has already begun since July 2017, having initiated a policy on AI development with the publication of the national artificial intelligence strategy called the 'Next Generation Artificial Intelligence Development Plan[12]'.

Recall also the road map outlined by the Chinese government from 2020 to 2030: it has set itself the goal of being competitive with other global competitors in the area of AI development, and then catching up with the United States in at least some areas of artificial intelligence until 2030, the year in which it wants to be the main center for innovation in this new technology.[13]

The United States, in order to curb Chinese dual capabilities and skills, had, as early as April 2023, through the mouth of Joe Biden's National Security Advisor, Jake Sullivan, stated that it wanted to protect US technologies "with a small backyard and a high fence [2][14]".

In spite of everything, the American willingness to make a paradigm shift remains unaltered, even while producing a radically new policy framework. This proves that

---

[10] On this point see H. KISSINGER, *Diplomacy*, New York 1994, Simon & Schuster, pp. 29–55.

[11] Cf. A. PAGANI, *Manuale di intelligence e servizi segreti*, Rubbettino, pp. 393 ff.; the author recalls another great volume on the point in the context of perspectives of war other than war proper, i.e. L. Qiao, X. Wang, "*War Without Limits. L'arte della guerra asimmetrica fra terrorismo e globalizzazione*", Libreria Editrice Goriziana, Gorizia 2001.

[12] See https://digichina.stanford.edu/work/full-translation-chinas-new-generation-artificial-intelligence-development-plan2017/.

[13] For more perspective on law, S. GUERRA, *L'intelligenza artificiale tra sperimentazioni normative e limiti etici nel mercato globale*, in *Riv. sci. della com. e di arg. giur.*—anno. XIV 2/2022, p. 105 ff.

[14] A. ARESU, *Beijing does not give up in the chip war*, in *Limes*, '*China remains a mystery*', No. 4, 2023, pp. 93–104.

all federal programs and regulations concerning technology are being reoriented towards a single goal: not to lose the competition with the People's Republic of China.

In fact, the regulatory path that the United States of America is following on the subject of AI is quite peculiar. Well, the litmus test can be seen in the US approach to the subject, which has reduced regulatory requirements to a minimum so as not to hinder companies that are working on sustainable technological development [1].[15] Just as the rise of the dragon in the field of technological supremacy is undeniably fierce and determined.

Well, in all of this, there is and could be room to appropriately summarise problems and solutions to this gigantic rivalry.

At the moment, however, the red line that must not be crossed is Taiwan, on which situations of serious conflict bordering on war would be triggered (notwithstanding the fact that the dossier on the island's hegemony is not a priority for either of them as far as long-term prospects and imperial projection are concerned, apart from the US-flagged destroyer Halsey).

The systemic characteristics of the two superpowers, for the well-known geopolitical and ideological reasons, de facto maintain a high level of competitiveness, with the consequent intentions of technological supremacy, which gives little hope for a 'good domestic AI society[16]'.

Nonetheless, one must also dwell on the repercussions that this technological challenge has on the global geopolitical equilibrium, as well as its exercise; on the other hand, the geopolitical exercise obliges one to reason, educating one on the limits not to be overstepped.

On the horizon could be the possible conclusion of the American century, aware of its isolationist posture, to make way for the much more resilient and enterprising Chinese century, in the terrain of AI technological conquest. And yet on the horizon, the joint idea of uniting to divide the globe is not peregrine. Maximum profit with minimum effort. It takes two to tango.

---

[15] S. ACETO DI CAPRIGLIA, *Artificial intelligence: a global challenge between risks, prospects and responsibilities. Le soluzioni assunte dai governi unionale, statunitense e sinico. Uno studio comparato*, in Federalismi.it, 09/2024, pp. 26 ff.

[16] Cf. S. ACETO DI CAPRIGLIA, *op.cit.*, pp. 26 ff.; see also E. HINE and L. FLORIDI, *Artificial intelligence with American values and Chinese characteristics: a comparative analysis of American and Chinese governmental AI policies*, in *AI & Society, 2022*, available at https://doi.org/10.1007/s00146-022-01499-8; for an accurate examination of artificial intelligence with regard to the concept of 'agere sine intelligere' see L. FLORIDI and F. CABITZA, Artificial intelligence the use of new machines, martini lecture, Bompiani publisher, pp. 115 ff.

7 The Competition Is on New Technologies

One of the main causes of the technology confrontation between China and America has centered on the 5G network without which many applications of artificial intelligence have no chance of functioning, think for instance of autonomous cars, which with low-power networks cannot function.

China has set its policy in the field of artificial intelligence on two State Council programs in particular: the Made in China 2025 and the Thirteenth Five-Year Plan for the Development of Strategic Emerging Industries. The two programs look with interest at AI on many pivotal points ranging from robotics to information technology, from electric cars to aerospace equipment. Beijing embraced it in 2017, when the State Council announced the Next Generation Artificial Intelligence Development Plan, which is important in two respects: increasing China's international competitiveness and defending national security.[17] China looks at AI with great attention, in fact, according to the Beijing authorities, it could lead to a 26% increase in GDP by 2030 and a 12% increase in the employment rate over the next 20 years. The Chinese government is aiming at two fundamental objectives for the strengthening of the country, which is going through a shaky economy determined in particular by the speculative bubble in the real estate sector, economic growth and social governance. To achieve these objectives, the development of the smart economy, of which artificial intelligence is a pivotal element alongside other technologies such as 5G, cloud computing, big data, the Internet of Things, edge computing, blockchain, hybrid reality and quantum computing. The integration of artificial intelligence into the economic circuit will radically change the socio-economic structure, production, labour, social culture and security models in the country of the Dragon. Hence, a new revolution in the wake of technology and innovation [6, 10, 11].[18]

Fundamental is the memory of events, in this key it must be remembered and certainly the US has not forgotten this as well as Russia, that the US won the Cold War among other things because it was able to dominate the technology sector, it was able to overcome the innovative capacity of the USSR. The most important case today is that of semiconductors, where the vast majority of companies were initially not even aware of the changing impact that microchips would have on their production. In the 1980s and 1990s, it was digital innovations that pushed the US economy out of the doldrums and triggered what has been called the digital revolution. Top-level productivity, globalisation and digitalisation determined the momentum of the US economy and geopolitical power in its heyday, that is, until 2008, the year of the financial crisis, a year that opened a rift in which the competition for dominance in artificial intelligence became part of a much bigger confrontation than just the

[17] "The State Council Program for Developing the New Generation of AI", State Council of the People's Republic of China, 2017.

[18] Limes n° 12/2022 Intelligence is not artificial (The US plan to slow down China's economy) by Paul Triolo.

(so to speak) fourth industrial revolution, but which will determine who will be the dominant power globally.[19]

References

1. Aceto di Capriglia S (2024) Artificial intelligence: a global challenge between risks, prospects and responsibilities. Le soluzioni assunte dai governi unionale, statunitense e sinico. Uno studio comparato. In: Federalismi.it, 09/2024, pp 26 ff.
2. Aresu (2023) Beijing does not give up in the chip war. In: Limes, 'China remains a mystery', no 4, pp 93–104
3. Cuomo C, Marino D (2019) The LOCAL WORK PLANS (LWP) and territorial economic system (TES): assessment and evaluation. In: Studies in systems, decision and control, vol 180, pp 225–237. ISSN: 2198-4182. https://doi.org/10.1007/978-3-030-00677-8_19
4. Hill C (2008) International business, economics and international strategy: the enterprise in global markets. Hoepli Ed.
5. Marino D, Tebala D (2021) Economy and development in Calabria: the weak development hypothesis. Riv Int Sci Soc anno CXXIX 2:187–220. ISSN: 0035-676X. https://doi.org/10.26350/000518_000068
6. Marino D, Gil Lafuente J, Tebala D (2023) Innovations and development of artificial intelligence in Europe: some empirical evidences. Eur J Manag Bus Econ 32(5):620–636. https://doi.org/10.1108/EJMBE-03-2023-0085
7. Marino D, Tebala D (2022) Rural areas and well-being in EU countries + UK: a taxonomy and a cluster analysis. Sustainability 14:15213. https://doi.org/10.3390/su142215213
8. Marino D, Timpano F, Gil Lafuente J (2023) The competition between legal and illegal firms in the market: theoretical models and empirical evidence. Eur Res Manag Bus Econ 29(3):100228. ISSN: 2444-8834. https://doi.org/10.1016/j.iedeen.2023.100228
9. Marino D, Cananzi D, Aragona F (2024) Artificial intelligence and ethics challenges of technologies. In: Ethics and artificial intelligence. Studies in systems, decision and control, vol 519. Springer, Cham. https://doi.org/10.1007/978-3-031-50902-5_3
10. Marino D, Trapasso R (2009) The new approach to regional economics dynamics: path dependence and spatial self-reinforcing mechanisms. In: Fratesi U, Senn L (eds) Growth and innovation of competitive regions. Advances in spatial science. Springer, Berlin, Heidelberg. https://doi.org/10.1007/978-3-540-70924-4_15
11. Marino D, Trapasso R (2020) Servitization and territorial self reinforcing mechanisms: a new approach to regional competitiveness. Investig Reg J Reg Res 48:85–93. https://doi.org/10.38191/iirr-jorr.20.022
12. O'Neill J (2011) The growth map: economic opportunity in the BRICs and beyond. Penguin, New York
13. Pagani. Manuale di intelligence e servizi segreti. Rubbettino, pp 393 ff.; the author recalls another great volume on the point in the context of perspectives of war other than war proper, i.e. Qiao L, Wang X (2001) War without limits. L'arte della guerra asimmetrica fra terrorismo e globalizzazione. Libreria Editrice Goriziana, Gorizia
14. Tebala D, Marino D, Bianchino A (2023) A provincial indicator to describe potential artificial intelligence. In: Marino D, Monaca M (eds) Artificial intelligence and economics: the key to the future. Lecture notes in networks and systems, vol 523. Springer, Cham. https://doi.org/10.1007/978-3-031-14605-3_1

[19] Limes n° 12/2022 Intelligence is not artificial (The real stakes of artificial intelligence) by Jacob L. Shapiro.

Artificial Intelligence Enables Precision Accounting Design and Tools

Demetrio Naccari Carlizzi, A. Calarco, F. Derrigo, and L. Marino

Abstract The use of AI enables a new heuristic capacity for public authority budgets that we call *precision.* Indeed, AI technologies make it possible, on the basis of the application of intelligent algorithms and prediction techniques, to construct an economic planning that reduces the *implementation deficit* of the planning process, bringing forecast and actual result closer together. The increased speed and accuracy of the budgeting process (near-real-time construction, evaluation and reprogramming) is generated by the data-driven model that rewrites programming documents and enables precise, customised and adaptive public policies. AI technologies are not exempt from risks to democracy and the legal principles underpinning public action, but they can prove to be valuable allies of law if the new *algorithmic public capacity* is translated into a structured and transparent by design framework that guarantees algorithmic legality and aims to give effect to legislative and meta-legal clauses that have hitherto been difficult to translate economically and administratively. In the same way, AI technologies should not be limited to supporting the political decision-maker, but can enable the development of the level of accountability of public action that fosters a participatory process and a model of *shared administration.* This new perspective paves the way for an unprecedented modelling of economic planning and the entire administrative ecosystem. The paper explores the hypothesis of a *precision economic planning* based on the use of AI that favours the ex-ante understanding of the impacts of local authorities' budgetary policies on the different generational cohorts. *Precision accounting*, an evolution of the traditional discipline, to which this work ideally tends, can offer a new administrative capacity at the service of programming that is not resolved in mere ideological choices or accounting labels but aims at an effectiveness of programmes and their impact, constructed data-driven and recalibrated in a continuous feedback. The objective is to change the government process through the support of data and artificial intelligence, democratising the budget and thus administrative action, making the decision-making process transparent and the administrative mandate assessable, thus consolidating the principle elaborated by the

D. Naccari Carlizzi (✉) · A. Calarco · F. Derrigo · L. Marino
Prepare for Change, 89100 Reggio Calabria, Italy
e-mail: demetrionaccaricarlizzi@gmail.com

D. Marino and M. Monaca (eds.), *Generative Artificial Intelligence and Fifth Industrial Revolution*, Lecture Notes in Networks and Systems 880,
https://doi.org/10.1007/978-3-031-73880-7_8

constitutional justice of the *Budget Public Good*, intended to synthesise, through a transparent allocation of resources, democratic choices.

Keywords Smart budgeting · Precision accounting · Generational accounting · Dynamic budgeting · Artificial intelligence in public sector

1 Background and Literature

The question posed in this paper is aimed at understanding the ability of AI to support government through a faster, transparent, fairer and more *accurate* budgeting process. This section of this paper reports on some of the *public accounting* issues debated in local government that stimulated the research question, in relation to the implications of using AI from the findings of the relevant scientific literature. In Sect. 2, the economic planning cycle of Local Governments and that of the State are analysed, in Sect. 3, an attempt is made to understand what impact AI can have on the public finance calendar, Sect. 4 outlines the support of AI to achieve *precision accounting* applied to the intergenerational impacts of planning, and Sect. 5 delivers the conclusions.

The fiscal crisis of States and the impact of cyclical financial crises also in local government, with the consequent creation of significant debt stocks and the passing on of even *implicit* debt to future generations, have prompted the EU in the recent past to impose new rules on *budget balancing on* individual States.[1] The Italian Parliament, in the same period, approved some controversial reforms such as the institute of *predestination*[2] together with a rethinking of the accounting model for local authorities, providing for an *accounting harmonisation* with the transition from an essentially financial accounting, through the so-called *enhanced financial accrual*,[3] to an *accrual accounting*, based on economic and equity principles [1]. This reform path is still in progress through various legislative interventions that have determined a rather tortuous framework, bureaucratising the economic planning cycle, overlapping fulfilments and instruments and in fact making it even more inefficient. Streamlining the budget process is, however, a need felt by administrators who are forced to approve a cumbersome succession of documents and allocate considerable resources to provisions (FCDE fund for doubtful debts) that stiffen budgets. Citizens

---

[1] With Constitutional Law No. 1 of 20 April 2012 on the 'Introduction of a balanced budget into the Constitutional Charter'.

[2] Articles 243-bis et seq. of Legislative Decree 267/2000 (TUEL), introduced by Legislative Decree 174/2012, introduced into Italian law the Predissertazione institute, which provides for a multi-year financial rebalancing procedure.

[3] Legislative Decree No. 118 of 23/06/2011 containing 'Provisions on the harmonisation of the accounting systems and balance sheet layouts of the regions, local authorities and their bodies, pursuant to Articles 1 and 2 of Law No. 42 of 5 May 2009' redefined the accounting system of local authorities in order to achieve homogeneity of budgets and accounts and to extend asset accounting to all entities.

themselves would be blessed by a dynamic management of resources and have the right to transparency of allocation choices for a *democratic* reading of the Budget. The right to a transparent and intelligible budget process as the basis for the exercise of democratic choice and the evaluation of the political-administrative mandate, as repeatedly emphasised by constitutional justice, which conceptually reconstructed the Budget as a *Public Good* [2] *functional to synthesising and rendering certain the choices of the territorial entity, both with regard to the acquisition of revenue and to the identification of the measures to implement public policies, an imperative duty for those who are called upon to administer a given community and to submit to the final judgement concerning the comparison between what is planned and what has been achieved*.

Administrators, on the other hand, need a planning framework that does not turn out to be a sequence of didactic fulfilments and that recomposes the current divarication between *accrual* and *cash in* order to achieve public budgets oriented towards *budget balances* rather than mere expenditure *coverage*. In other words, to the macroeconomic necessity of a dynamic allocation of resources in the expenditure items of the budget to respond to economic events [3] we need to add a dynamic cash management and an adaptive and reactive dynamism of policies to achieve and calibrate precision results to reconcile the public sector with the speed imposed by technology [4].

These just mentioned represent some of the issues that, in the field of local administration and in Public Accounting Law, can be defined as open and with respect to which the use of AI can make a paradigmatic contribution towards a greater capacity to give effectiveness to forecasts and efficacy to public policies. The problem of the need for a dynamic management of resources arises, in fact, also as a consequence of the ever-increasing rationing of resources from the central level to the territorial levels, but it also represents a challenge to revive public action and reduce the implementation deficit of policies. In this sense, there is a tendency on the part of legislators, supranational governing and representative bodies, and organs of justice, such as the Constitutional Court, to increasingly question themselves on the problem of the effectiveness of regulatory provisions, on the real enforceability of services defined with rather indeterminate legal and meta-legal clauses[4] on the need to measure the objectives of institutions at the various levels[5] that without an adequate capacity for

[4] Letter m of the second paragraph of Article 117 of the Italian Constitution establishes the exclusive competence of the state on the 'determination of the essential levels of the services concerning civil and social rights that must be guaranteed throughout the national territory'. The OEL clause has remained an enigmatic definition without an actual quantification. The delegated Law No. 42 of 2009 on Fiscal Federalism provides for some parameters defined as standard requirement and standard cost to anchor the financing of the functions of municipalities, metropolitan cities and provinces and for the financing of the essential functions of the regions (health, education, assistance) for the definitive overcoming of historical expenditure.

[5] ISTAT's Sustainable and Fair Welfare project provides a set of 152 indicators for measuring well-being reported annually in the BES Report. The 2030 Agenda is a programme of action for sustainable development signed by 193 UN member countries in which 17 Sustainable Development Goals (SDGs) divided into 169 targets are identified.

quantification and forecasting risk appearing enigmatic as in the case of the identification of the Essential Levels of Benefits [5] and reveal an insufficient and unbiased allocation of resources, unbalancing the budgets of territorial authorities. The problem of intergenerational equity, as we shall see in Sect. 4, is paradigmatic of the need to make the public budget immediately readable and to strengthen the heuristic capacity of accounting tools, in the face of the problem of the fiscal crisis in the public sector and the need to guarantee the principle of equality in a diachronic sense towards new and future generations, consistent with the principles of constitutional jurisprudence [6].

Studies on how AI can improve the process that oversees public budgeting, defining an effective transition from *e-budgeting* to *smart budgeting*, are limited and recent. Empirical evidence already exists that the use of Big Data in budgeting can increase efficiency and effectiveness while reducing costs [7]. Smart budgeting is seen as a systematic process that collects relevant information and uses algorithmic models to develop a budget and produce scenarios to support decision-making [8]. The budget cycle (planning, programming, budgeting, appropriations, control and evaluation of financial resources) is being studied towards a transformation to e-budgeting [9] through digitisation of budget procedures and the dissemination of open and Big Data [10]. The doctrine valorises the role of e-budgeting in contributing to efficient budgets and improving *governance* as budget processes provide quantities of data that can enable the detection of patterns, which can be used to design more efficient and effective ways to build a budget for a policy [11]. In this sense, e-budgeting, if integrated with financial management, can lead to better *governance* [12]. AI can operate as a decision support using statistical and machine learning techniques, some authors have highlighted the relevance of the organisational context for budget functions, using text mining tools [13]. The use of artificial intelligence techniques, as a tool for data analysis, has the potential to support better decision-making in government. Artificial intelligence should not necessarily be used to automatically make final decisions, but to provide ideas and identify potential opportunities and different scenarios that government leaders can then evaluate and incorporate into their decision-making processes [14]. There are experiments that have involved *machine learning* algorithms for predicting disruptions in municipalities using data from the Ministry of the Interior and financial indicators from ISTAT [15]. Other authors have explored artificial intelligence techniques to help allocate public spending, using *multi-objective genetic algorithms*, proposing inputs that are processed (public budget expenditures) to generate outputs (economic, political and social outcomes) [16]. Experiments on the use of smart algorithms indicate a transition process towards Smart Budgeting also in the public sector, aiming at the mission of the sector, which is the generation of public value. AI and *smartness* are also seen as factors in organisational implementation. Smart technologies have the potential to foster the co-creation of public services and the generation of public value in management processes based on their collaborative, social and horizontal nature [17]. The current reality of public finance management despite significant investments in IT shows how the prevailing paradigm has struggled to achieve a successful digital transformation, due to an inadequate financing model and a closed, unresponsive and inflexible technological architecture that, instead, would need openness to build an ecosystem of infrastructure, data and shared services, putting data at the basis of the *digital transformation* not only of public finance but of the entire governance process [18]. The knowledge of data held by public administrations, processed by AI algorithms, could be a valuable aid in making budgets transparent to citizens, both in understanding ex ante the real beneficiaries and in detail the benefits of a public policy. AI could contribute to solving the problem of information asymmetry between the public administration and citizens and offer the policy-maker and the citizen, in a participatory process, possible options on which to make an informed choice [4] and act as an enabler, in an *open government* context, of administrative transparency as a basis for participation and subsidiarity processes. In the same way AI could solve the issues of resource tracking in cases of expenditure constraints and allocation of public expenditure aimed at accounting or constitutional principles (capital expenditure/current expenditure, priority allocation of resources on expenditure items linked to essential levels of performance, extra LEP expenditure) or vertical allocation between levels of government. Precision accounting would be attentive to the reasons of equity and sustainability as an accounting qualification of *algorithmic public capacity* [4].

2 Programming Cycle in Local Authorities

In this context, characterised by a multiplicity and fragmented nature of experiments and acquisitions, there emerges the need to rethink the tools and design of public economic-financial programming, on the one hand in order to prepare a model suitable to act as an ecosystem for the use of data and algorithms, and on the other hand to temporally connect the action of public administrations with the time imposed by technology and pave the way for accounting with an increased descriptive and administrative capacity. Intelligent allocation and management of resources would allow near real time governance of the planning cycle and the budgetary process, enabling dynamic precision calibration of public policies while measuring and verifying the generation of public value. This is due to the predictive capacity of AI to support the design of public policies and their implementation, which reduces their traditional imprecision [4] i.e. being controversial in formulation, uncertain in effectiveness, fragile and random in results [19]. The design of economic planning at various levels suffers from this uncertainty and in an attempt to rationalise the process has imposed an overly theoretical, complex and cumbersome planning. The sequence of documents is temporally misaligned in order to reconcile the previous year's operating results to draw up an authorising budget for the following year, with some verification steps during the year, purely on coverage and trend reconciliation, while a true reconciliation between forecasts, commitments made and the expenditure actually made takes place after the half-year of the following year, at the time of the statement of accounts. The feedback on the effectiveness of the policy comes at the end of this long process, therefore in the middle of the following year and when planning for the new year is already well under way. This scheme betrays a merely ideological and conceptual character of public policies, constructed without an adequate data base, monitored ex post according to the traditional scheme of the public policy cycle, constantly lagging behind the needs of social dynamics. In Italy, where politics and the media devote more attention to political factors than to policies [19] and government action is assessed on legal and financial parameters rather than on the result, this translates into planning that is very sensitive to political propaganda and less to the real effectiveness of measures.

3 Programming of Local Authorities with Artificial Intelligence

The sequence of economic planning in local government thus hands us a scheme so cumbersome and burdened with mandatory documents as to mark a constant delay in the flow of data from the *budget* to the *statement of accounts* (which takes place in the six months following the reference year) as shown in Fig. 1.

Planning is defined as a *circuit, a process by progressive updates whose objective is to lead towards a path of value, qualitative-quantitative and financial coherence through which the contents of the entity's will are formalised* [20]. The consistency of the various instruments can be found in an internal bureaucratic and *initiatory* logic rather than in an actual functionality of the fulfilments. In the intent of the legislator who first introduced the Executive Management Plan (Article 169 TUEL) and more recently the Single Programming Document (DUP) and the Integrated Activity and Organisation Plan (PIAO), there was a *shift from a season of strategic and operational programming tools that were completely, or almost completely, divorced from the economic-financial reality of the authority and the public finance context, to a new approach to local planning that makes the programming process a useful and necessary prerequisite for planning activity* [21]. The DUP is the instrument that should translate the planning lines and would be the necessary prerequisite for all

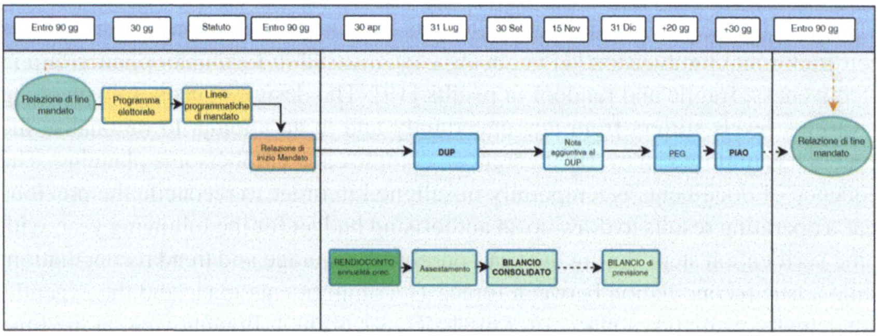

Fig. 1 Flowchart on a timeline of a process for a mandate cycle

other planning documents. It consists of two sections: the strategic section with a 5-year time horizon and the operational section that has a programmatic and financial content and contains the operational lines of the entity's planning. The latter supports the forecasting process for the preparation of the budget, defining missions, programmes and objectives to be achieved. The PIAO is intended by the legislator to be the tool to ensure the quality and transparency of administrative activities and to improve the quality of services through a continuous process of simplification and reengineering of processes. The municipal executive body approves it for a three-year period and, according to the logic of *rolling administration*, it is updated annually. It integrates all the elements of sectorial planning and defines the programmatic and strategic performance objectives, the human capital management and organisational development strategy, the tools and objectives for recruiting new resources and enhancing the value of internal resources, the tools to achieve full transparency, the list of procedures to be simplified and re-engineered, and the methods and actions aimed at full compliance with gender equality. The PIAO would have the ambition of becoming a unitary, integrated tool that replaces the plans of the past and metabolises them into a new and all-inclusive, *cross-cutting* instrument that allows a 360° analysis of the administration and all its objectives to be planned [22]. The Executive Management Plan (PEG), adopted by the municipal council within twenty days of the approval of the budget, details the operational planning contained in the DUP. It should link the executive body of the municipality and the heads of the services in the implementation of the programmes and define the management objectives and the results expected from the implementation of the programmes, simultaneously allocating the resources necessary for their pursuit. General, or first-level, objectives are defined in the PEG, while executive, or second-level, objectives are identified in the Detailed Plan of Objectives and the Performance Plan, which are part of the PIAO. In day-to-day administration, these instruments run the risk of being perceived as rhetorical fulfilments that only claim a result because the law assigns them that task. The various documents follow one another in time and in some cases overlap because they are the result of a doctrinal logic that is somewhat divorced from the concrete experience of an entity's operations. The Council of State intervened with sentence No. 4426 of 17 May 2024, contradicting the previous decision of the Apulia Regional Administrative Tribunal No. 256 of 7 February 2023 that had annulled, on the appeal of some municipal councillors, the resolutions approving the DUP and the outline of the budget estimate that took place in the same session. If the logical and chronological sequence of the two accounting documents can be resolved in a mere order of approval, this shows that the function of the SPD is reduced to a mere formal and descriptive fulfilment that imposes on the subsequent documents a mere tendential consistency that greatly reduces the substantial meaning of the overall design of the planning cycle. Again, the Council of State, in its opinion 922 delivered on 22 May 2022, states, verbatim, that "...the risk that the SPD may be resolved into a juxtaposition of plans, almost defining itself as a further layer of bureaucracy, does not appear to have been dispelled", "...having unified in a single instrument several plans already provided for under current legislation highlights, on the other hand, the duplication or, in any case, the uncertain boundaries of the planning activities to which public administrations are called". On closer inspection, it is the entire cycle that is configured as a sum of *layers of bureaucracy, which is* extremely difficult to recognise as a sequence of effective tools for defining a rapid and functional planning framework. It is no coincidence that every year the Ministry of the Interior intervenes to postpone the deadline for the approval of the budget for municipalities, in some cases allowing them to operate through the institution of the provisional financial year for more than half a year, thus demonstrating the fragility and ineffectiveness of the planning calendar and the documents and tools provided. To avoid this juxtaposition of plans and redefine the programming cycle in the light of the use of AI techniques, we can depict in Fig. 2 a new framework representing a possible sequence towards the definition of an e-cycle: the programming e-cycle.

The main effect would be the contraction of phases, which would reduce the time of the planning and budget management cycle. It would become a continuous and rapid process, tending to approximate accrual and cash criteria. The end-of-term report would be reconciled at 31/12 of the last year (or the last day of government) together with the statement of accounts. This would also provide a certain data base to verify the congruity of the programmes proposed by mayoral candidates during the election campaign and of the elected mayor when communicating his or her mandate programme lines, establishing a close link with the end-of-term report and instruments such as the DUP and PIAO.

The DUP and PIAO could be replaced by a single planning document, constantly and dynamically updated, offering procedural simplification but also a real representation, with an adequate quantitative and real time analysis of the processes in place. All this would allow, in relation to the real trend of revenue and expenditure, a monitoring of the timetable and its continuous recalibration. In this sense, the additional note to the Dup would no longer make sense. Budget and Peg, thanks to AI technologies, could constitute one and the same accounting document, using in particular predictive algorithms. The *adjustment* would be a fulfilment absorbed in the dynamic and real-time process as well as the *consolidated budget of the entity.* The *statement of accounts* would no longer be necessary as an autonomous phase because it would be incorporated into a dynamic and continuous phase thanks to the continuous inflow of data and the digitisation of services and offices. The existing spread between forecasting and reporting in terms of deficit implementation would in any case be greatly reduced. It would radically change the entire budget cycle, which today rests on the *end-of-term report* of the previous administration and is based on the *mandate programme* approved in the Council, and then structured with instruments such as the DUP and the PIAO, which clearly separate the strategic and operational phases. This revision of processes would enable channels of participation for citizens and stakeholders called to active participation, in a model of shared administration facilitated by the opening up of data, the reduction of the information asymmetry between PA and citizens [23] by techniques of analysis and visualisation of big data [24] to improve decision-making processes using electronic democracy, to support decision-making in the public sector based on the use of artificial intelligence (AI) and big data.

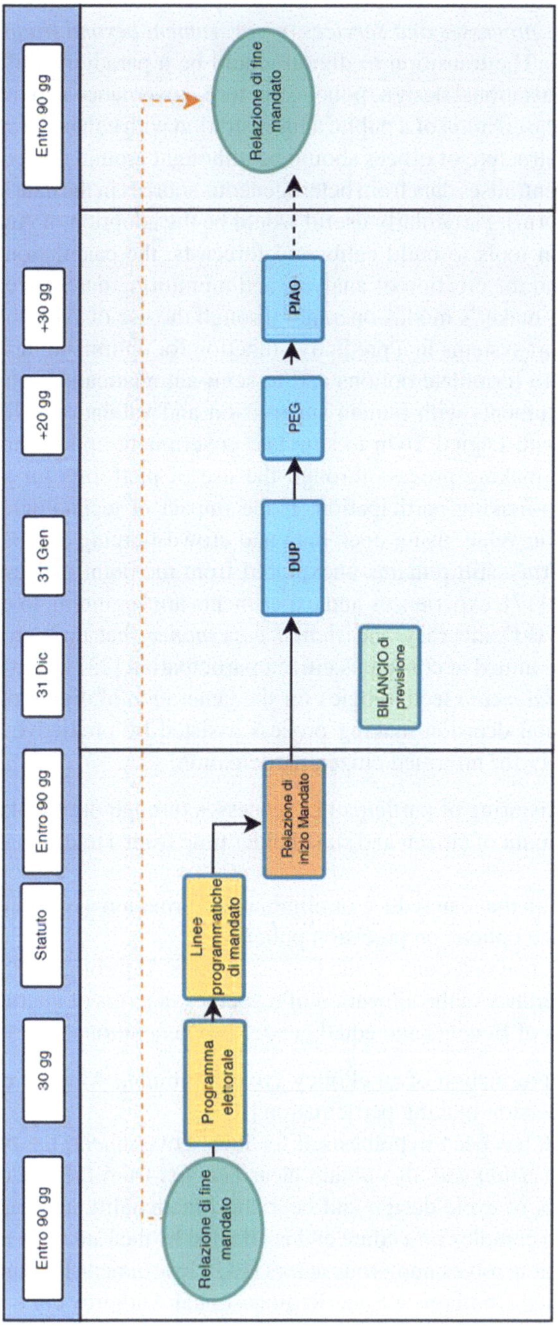

Fig. 2 Flowchart of a mandate process timeline

Smart budgeting assumes not a simple digitisation but a complete digital transformation of the budgeting process and of the entity understood as *a holistic effort to overhaul the core processes and services of government beyond traditional digitisation efforts* [18]. The transition to digital would be a paradigm shift involving an overhaul of organisational design, policy structure, governance and management processes and thus the creation of a public administration with enhanced capabilities. The organisational structure of offices should be rethought around the technological infrastructure that centralises data from heterogeneous sources in the data lake (Smart City Big Data platform). Particularly useful would be the adoption of Analytics and Infodatavisualisation tools to build calibrated forecasts, the calculation of performance indicators and the creation of analysis and monitoring dashboards. It would change the decision maker's modus operandi through the use of AI techniques and DSS decision support systems in a predictive function for optimising resource allocation and Gen AI to formulate options and to semi-automatically prepare reports and accounting documents with human supervision and validation. The authority could equip itself with Digital Twin to simulate government options and simplify the public decision-making process through the use of platforms for stakeholder and citizen decision-making participation. If the impact of technologies to foster the creation of public value, using open data and crowdsourcing and participatory websites and platforms, still remains unexplored from the point of view of public sector management [17], experiences and experiments are beginning to emerge that point to models of E-Democracy and *shared governance* that through the use of digital platforms are aimed at conscious citizen participation [23] and at harnessing the potential of *social media* technologies for the generation of digital public value [25]. Indeed, a digital decision-making process assisted by predictive algorithms offers the opportunity for informed citizen participation:

1. enabling the structuring of participatory processes through digital platforms;
2. reducing the amount of citizen and stakeholder time from a traditional participation process;
3. offering open data that can reduce or eliminate information asymmetry;
4. submitting choice options on precision policies;
5. indicating predictive outcomes of the budget options AI through DSS can offer;
6. ensuring transparency in the allocation of resources in terms of guaranteeing the Essential Levels of Benefits and equal gender and generational opportunities.

Figure 3 is a representation of an ePolicy cycle combining AI as an opportunity for active citizen decision-making participation [23].

Similarly to what has been hypothesised for local government, the timetable of the State *Budget Decision* can also obtain clear benefits from the use of AI techniques both in terms of cycle design and heuristic functionality and capacity. The *Budget Decision* is a complex procedure and is affected by the European governance of public finance that involves numerous actors (EU-Government-Parliament-Local Authority and Unified Conference State-Regions-Local Authority) in a multi-level process (EU-State-Regions-Local Authority) that starts from the EU guidelines and is defined by the Parliament at the initiative of the Government. This dual propulsion

Fig. 3 Precision ePolicy Cycle—Choice of option Diagram

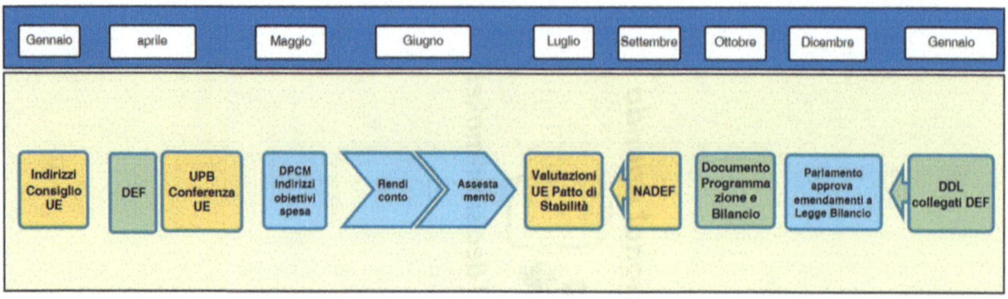

Fig. 4 National Budget Decision process

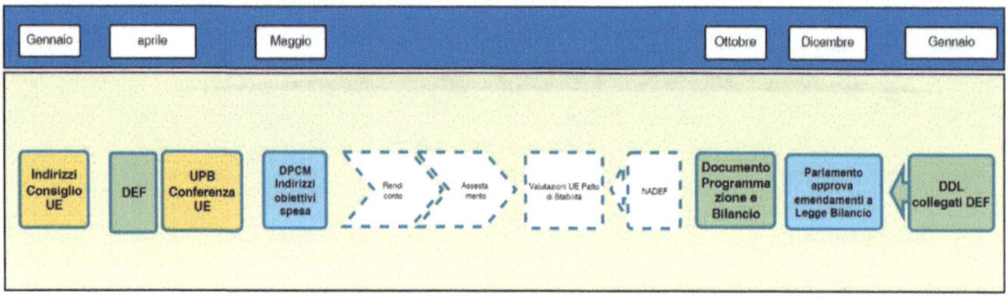

Fig. 5 National Budget Decision simplified process

(national line and European line) intersects between political guidance and sanctioning power and is based on allocative assessments and compliance with multi-year macroeconomic targets and forecasts. Figure 4 shows the Public Finance Calendar describing the process of the Budget Decision.

Figure 5 shows the public finance calendar in the light of the simplification implemented with the help of artificial intelligence.

Rendiconto, Assestamento and Nota di Aggiornamento al Documento di Economia e Finanza (NADEF), would be substantially absorbed in the forecast documents thanks to AI algorithms. The assessments of the European institutions concerning compliance with the parameters of the *Stability Pact* would take place near real time in the forecast thanks to predictive algorithms without significant subsequent deviations.

Schematically, the use of AI would make it possible to:

1. reduce the strictly political propaganda dimension of forecasts, evaluations and decisions by highlighting the real impact ex ante;
2. align forecasts and results and reduce the implementation deficit;
3. contribute to greater veracity, reliability and predictive capacity to the budget;
4. favouring choice between public policy options;
5. allocate resources more efficiently;
6. achieve results in terms of *public value*, exceeding the output threshold;
7. allow for evaluations during the public policy cycle to deal with unforeseen events or to recalibrate the intervention and restart the cycle;

AI techniques would lead to significant effects on accounting:

1. drastic simplification and reduction of accounting documents;
2. accrual-based and cash-based accounting approach;
3. trend reduction of *residual assets and liabilities*;
4. precision prediction of *outcome* and public value;
5. shifting the focus from the concept of *coverage* to that of *budgetary balance* by freeing up resources;
6. greater effectiveness of public decision-making in terms of the programmability of choices;
7. audit by the auditors and the Court of Auditors truly *concurrent*;
8. transparency of evaluation on the basis of completeness of data by the internal evaluation bodies (OIVs).

Similarly, we can estimate the effects on the PA model and its heuristic capacity. A data-driven decision-making process assisted by predictive algorithms would be truly rationalised, precisely because it would be based on solid information bases, effective evaluation and forecasting techniques and reality data, reducing information asymmetries, time lags and ideological or symbolic evaluations.

We would witness a democratisation of allocative choices and an effective enabling of the concept of the Public Good Budget [2] and of the constitutional meta-legal clause of the Essential Levels of Services[6] and of the financing criteria of the *Standard Cost* and *Requirement*. Accounting would become *precision*, allowing the role of the citizen as arbiter to finally be valorised (democratic principle). Administrative transparency would be fostered by the open availability of data and the constant reading of the progress of programmes, made available with the help of *dashboards*. All this leads to the acquisition of a new *algorithmic administrative capacity for PA precision*.

4 Artificial Intelligence and the Problem of Intergenerational Equity

One of the most jurisprudential problems is the assessment of intergenerational equity and sustainability [6, 26–31] in relation to the benefits of policies and the distribution of the related tax burdens on taxpayers. Some particularly alert scholars have raised the problem of the continuous production, behind apparently neutral conventional accounting labels, of problematic implicit debts passed on to subsequent generations as an unfair burden that would impoverish them. It is to the merit of the Italian Constitutional Court to have reconstructed, with a work of extraordinary merit in terms of

[6] Delegated Law No. 42 of 2009 on Fiscal Federalism provides for some parameters defined as Standard Requirements and Standard Costs to which to anchor the financing of the functions of municipalities, metropolitan cities and provinces and for the financing of the essential functions of the Regions (Health, Education, Assistance) for the definitive overcoming of historical expenditure.

doctrine and jurisprudence, the principle of intergenerational and infra-generational equity, also resolving the problem of the protection of these principles through the legitimacy of the Court of Auditors to raise questions of constitutional legitimacy when controlling the budgets and financial balances of local authorities. In some valuable decisions from 2019 onwards, with Judgments No. 4 and No. 18, the *Judge of Laws* has censured the *perpetuation of the structural deficit over time, through a regulatory drip of postponements, ends up paralysing any reasonable recovery project, thus colliding with both the principle of intragenerational and intergenerational equity*. Of particular significance for our purposes is the fact that the Court has intervened to contain the repayment of the debt produced within the rules and timeframes of the three-year planning of local authorities and, in any case, within the five-year term of the administrative mandate, in deference to the democratic principle of the mandate. Economic doctrine has used as a method of verifying the intergenerational equity of programmes that based on *generational accounts* [26]. This method suffers from a long-term horizon, which is useful for economic analysis that is mostly retrospective, but late for the impacts already produced on people's lives. The application of AI algorithms to public budgets, on the other hand, allows for unprecedented accuracy of analysis and dynamic monitoring of expenditure trends. These results, combined with the possibility of structuring precise and adaptive policies, can enable timely intervention by the decision-maker, control bodies (the Court of Auditors), stakeholders, and citizens in general, also to protect the interests of the younger generations and in favour of intragenerational fiscal equity. The use of AI techniques could ensure the dynamic tools capable of a diachronic reading of laws with a high financial content in order to give effectiveness and adequacy to the need pointed out by authoritative doctrine of a financial investigation on the evolution of the conception of fundamental rights as meta-legislative values, as axiological instances of control of the choices of positive law [32]. Principles such as the Balanced Budget, the Dynamic Budget Balance and relevant public finance issues such as the regulation of financial flows between State-Regions-Local Authorities and the macroeconomic projections of costs for the protection of the Essential Levels of Services can find in the use of artificial intelligence techniques effective financial management tools. If public budgets need to highlight 'mandate accounting' and the link between harmonisation of public accounts and accountability of the electoral mandate [33] the role of the principle of Transparency could be supported by artificial intelligence techniques that would have the task of making the accounting language intelligible to the citizen voter. To study the potential of a precision balance sheet, a working group (Demetrio Naccari Carlizzi, Antonio Calarco, Lorenzo Marino) was set up to reclassify the items of a municipality's balance sheet into macro-categories of generational interest and to study the formulas for calculating each typical expenditure according to the fiscal impact and benefits on the individual generational cohorts. With the challenge of precision accounting in the background, a further level of detail was defined in the Chart of Accounts, in addition to those provided for in Appendix no. 6/3 of Law No. 118/2021, to link financial accounting with precision accounting (impact on the generational cohort and impact on beneficiaries). The AI, suitably trained by means of training sets of budget data from the last 20 years, data from the

Territorial Information System (SIT), ISTAT and demographic data, and additional technical parameters, with reference for example to the expenditure on buildings for crèches, has elaborated

- the need for kindergartens in the City of Reggio Calabria;
- the optimal location of the facilities within the municipal territory;
- the related amount for financing the facilities;
- the multi-year breakdown for the construction of kindergartens taking into account the generational cohorts 0–3 years.

The generation impact (Ig) of an investment expenditure in relation to its life cycle can be calculated at a general level using the following formula:

$$\mathrm{Ig} = (\mathrm{Bg} * \mathrm{De1}) - (\mathrm{Og} * \mathrm{De2})$$

Bg Generational benefits: refers to the direct and indirect benefits that each generational cohort obtains from public spending, could include benefits such as education, health care, social services, job creation, infrastructure investments.
Og Costs for generation: represents the direct or indirect costs that a specific generation faces.
De1 Duration of beneficial effects on generation.
De2 Duration of charges on generation.

With this in mind, AI techniques allow a budget to become smart and in this case to indicate exactly the benefits and burdens of a generation and an individual member thereof, distinguishing between direct beneficiaries, secondary beneficiaries (family), indirect beneficiaries (social effects), and net contributors without direct benefit.

AI can contribute to intergenerational equity in a public budget through:

1. analysis of demographic data and trends to identify existing inequalities and predict future trends, enabling the government to adopt more targeted and balanced policies;
2. optimisation of resource allocation (using algorithms and predictive models to allocate resources optimally and more fairly, identifying implicit and indirect debts that are often hidden behind neutral formulas in accounting);
3. simulations and impact assessments: through its predictive capacity applied to expenditure;
4. promotion of intergenerational equity: AI, appropriately trained on quality data free of *bias*, can help to expose prejudicial choices and the passing on of debt to single generational cohorts;
5. continuous monitoring and evaluation: trained models can carry out real-time evaluation of policy developments and suggest course corrections;
6. support for the definition of intergenerational rebalancing policies;
7. evaluation of the outcome of individual policies on each generational cohort.

Predictive analysis and modelling using machine learning algorithms to analyse large amounts of historical and demographic data can be used to support policy makers in drawing up a more equitable balance between generations and simulate scenarios based on the policy options considered by policy makers.

5 Conclusions

The partial findings of the study suggest that artificial intelligence techniques offer greater transparency of financial investigations and the guarantee of greater reliability of annual budget and mandate planning forecasts. If law is in search of the effectiveness of administrative rules and decisions, of an *effective administrative law*, i.e. a *law that* is *observed, implemented, and harbinger of results consistent with the objectives for which it was envisaged* [34, 35], artificial intelligence can be a valuable ally because it increases the forecasting and implementation capacity of public administrations, thus ensuring the effectiveness and transparency of financial rules and public policy outcomes. AI technologies enable a new heuristic capacity in accounting and budget planning and improve the effectiveness of the government process, which becomes one of *precision*. To unleash the potential of AI, action must be taken to adapt the design of the economic planning cycle, reviewing and simplifying individual accounting documents, starting with the processes that AI enables. At the same time, local and national government must rethink its organisational structure and administrative processes to create a data-driven ecosystem through a digital transformation path that enables *data governance*. The legislation on the Digital Administration Code (CAD Legislative Decree No. 82 of 7 March 2005) provides regulatory rules and general guidelines, but the construction of the infrastructure that collects data, the structuring of digital technologies and AI are substantial processes that require a convinced adherence to the paradigm shift in government and decisive organisational changes. AI can revitalise the democratic process by truly enabling the Balance Sheet as a *public good, functional in synthesising and rendering certain the choices of the territorial body* as indicated by Constitutional Jurisprudence. In this sense, technology can support both administrative and accounting principles (*transparency, clarity and specification of budget lines*), and constitutional principles (intergenerational and intragenerational equality, democratic principle of mandate), as well as accounting rules and constraints (impartiality in the allocation and vertical distribution of resources, compulsory and incompressible expenses pursuant to art. 8, paragraph 1 of law 42/2009) and metalegislative clauses (Essential Levels of Services, standard cost, standard requirements) that with ordinary accounting risk being ensured only nominalistically. The review of the guarantee of the essential services to be guaranteed to the citizen and the traceability or otherwise to the constraint of Article 117, second paragraph, letter m (expenses that guarantee the LEP) needs an adequate financial investigation and continuous verification, substantial and traceable, that the smart algorithms of a smart budget would ensure. The reduction of public resources available to governments, with the consequent *colonisation of* the budget by irrepressible rights, militates for an increasingly effective use of resources and for the definition of adaptive and precision policies with a macroeconomic projection of public choices. In this sense, accounting, enlightened by AI, can guarantee a result-oriented budgeting process for administrative action and save some constitutional provisions from non-implementation, recovering the effectiveness of legal principles and the protection of citizens' rights.

References

1. Piano Nazionale di Ripresa e Resilienza. Riforma 1.15—Dotare le Pubbliche amministrazioni di un sistema unico di contabilità economico-patrimoniale accrual
2. Corte Costituzionale sentences n. 184/2016 e 247/2017 Bilancio Bene Pubblico
3. Valle-Cruz D et al (2022) From E-budgeting to smart budgeting: exploring the potential of artificial intelligence in government decision-making for resource allocation. Gov Inf Q 39(2). Article 101644
4. Carlizzi DN, Quattrone A (2023) Artificial intelligence and data governance for precision ePolicy cycle. In: Marino D, Monaca M (eds) Artificial intelligence and economics: the key to the future. Lecture notes in networks and systems, vol 523. Springer, Cham. https://doi.org/10.1007/978-3-031-14605-3_6
5. Comitato Tecnico Scientifico with investigative functions for the identification of LEP (CLEP) established by the decree of the Presidente del Consiglio dei Ministri, 23 Mar 2023
6. Sentences of the Corte Costituzionale n. 4/2019, 18/2019, 115/2020, 34/2021, 235/2021, 246/2021, 219/2022
7. Valle-Cruz D et al (2022) From E-budgeting to smart budgeting: exploring the potential of artificial intelligence in government decision-making for resource allocation. Gov Inf Q 39(2):101644. ISSN 0740-624X. https://doi.org/10.1016/j.giq.2021.101644. https://www.sciencedirect.com/science/article/pii/S0740624X21000800
8. Manyika J et al (2011) The next frontier for innovation, competition, and productivity. McKinsey Global Institute
9. Gil-Garcia JR et al (2014) Being smart: emerging technologies and innovation in the public sector. Gov Inf Q
10. Puron-Cid G, Ramon-Gil G (2004) Enacting e-budgeting in Mexico. Public Financ Manag 4:182–217
11. Sgueo G (2015) Electronic budgeting. Innovative approaches to budgeting. European Parliamentary Research Service
12. Höchtl J et al (2016) Big data in the policy cycle: policy decision making in the digital era. J Organ Comput Electron Commer 26(1–2):147–169. https://doi.org/10.1080/10919392.2015.1125187
13. Gamayuni R, Hendrawaty E (2020) E-planning, E-budgeting and the quality of government institution performance accountability system in Indonesia. Talent Dev Excell 12(1s):218–225
14. Anastasopoulos LJ et al (2020) Organizational context and budget orientations: a computational text analysis. Int Publ Manag J
15. Valle Cruz D et al (2020) Towards smarter public budgeting? Understanding the potential of artificial intelligence techniques to support decision making in government. In: The 21st annual international conference on digital government research, June 2020, pp 232–242. https://doi.org/10.1145/3396956.3396995
16. Franca S (2022) Il trattamento dei dati nelle sperimentazionidi intelligenza artificiale riguardanti le pubbliche amministrazioni. In: Paino A, Donati F, Perrucci A (eds) Intelligenza artificiale

e diritto: una rivoluzione, vol 2. Amministrazione, responsabilità, giurisdizione. Il Mulino, p 165

17. Criado JI, Gil-Garcia JR (2019) Creating public value through smart technologies and strategies: from digital services to artificial intelligence and beyond. Int J Public Sector Manag 32(5):438–450. https://doi.org/10.1108/IJPSM-07-2019-0178
18. Long C et al (2023) Digital public financial management: an emerging paradigm. ODI working paper. ODI, London. www.odi.org/en/publications/digitalpublic-financial-management-an-emerging-paradigm/
19. Bobbio L et al (2017) Le politiche pubbliche, problemi, soluzioni, incertezze, conflitti. Mondadori Università, Milano. ISBN-10: 8861844669. ISBN-13: 978-8861844667
20. Fiorillo G, Lombardi L (2023) Gli strumenti di programmazione finanziaria dell'Ente locale, con particolare riferimento al Piano della Performance. www.filodiritto.com
21. Delfino F (2021) I principi contabili e la programmazione di bilancio. In: Bilardo S, Anzalone M (eds) Manuale di contabilità delle regioni, degli enti locali e dei loro enti, organismi e società. Molfetta
22. Tascone I. Il nuovo Piao: modelli, obiettivi, contenuti. ASMEL
23. Naccari Carlizzi D et al (2023) Decision making and e-democracy: Archimede, a tool to support the data-driven process. In: Marino D, Monaca MA (eds) Innovations and economic and social changes due to artificial intelligence: the state of the art. Studies in systems, decision and control, vol 222. Springer, Cham. https://doi.org/10.1007/978-3-031-33461-0_5
24. Matheus R et al (2020) Data science empowering the public: data-driven dashboards for transparent and accountable decision-making in smart cities. Gov Inf Q 37(3). ISSN 0740-624X. https://doi.org/10.1016/j.giq.2018.01.006
25. Zavattaro S, Brainard L (2019) Social media as micro-encounters: millennial preferences as moderators of digital public value creation. Int J Public Sect Manag. https://doi.org/10.1108/IJPSM-02-2018-0059(2019)
26. Auerbach AJ et al (1991) Generational accounts—a meaningful alternative to deficit accounting. Working paper no. 3589. National Bureau of Economic Research, Cambridge, MA
27. Bifulco R et al (2008) Un diritto per il future. Teorie e modelli dello sviluppo sostenibile e della responsabilità intergenerazionale. Jovene editore
28. Monti L (2013) Teoria e principi di politica economica intergenerazionale. LUISS Academy
29. Naccari Carlizzi D (2009) Il bilancio di genere, la Calabria diventa una regione sostenibile. In: Ciarletta MS (ed) Il bilancio di genere in Calabria: strumento di sviluppo sociale e territoriale. Rubettino editore. ISBN 884982596X
30. Marino D, Trapasso R (2009) The new approach to regional economics dynamics: path dependence and spatial self-reinforcing mechanisms. In: Fratesi U, Senn L (eds) Growth and innovation of competitive regions. Advances in spatial science. Springer, Berlin, Heidelberg. https://doi.org/10.1007/978-3-540-70924-4_15
31. Marino D, Trapasso R (2020) Servitization and territorial self reinforcing mechanisms: a new approach to regional competitiveness. Investig Reg—J Reg Res 48:85–93. https://doi.org/10.38191/iirr-jorr.20.022
32. Carosi A (2018) La Corte Costituzionale tra autonomie territoriali, coordinamento finanziario e garanzia dei diritti. In: Manganaro F (ed) Scienza delle pubbliche amministrazioni. Editoriale Scientifica. ISBN 978-88-9391-269-3
33. Carosi A (2018) Il principio di Trasparenza nei Conti Pubblici. In: Rivista AIC n. 3/2018
34. Corso G et al (2022) Diritto amministrativo effettivo, una introduzione. Il Mulino. ISBN 978-88-15-29386-2
35. Marino D, Gil Lafuente J, Tebala D (2023) Innovations and development of artificial intelligence in Europe: some empirical evidences. Eur J Manag Bus Econ 32(5):620–636. https://doi.org/10.1108/EJMBE-03-2023-0085

Who's Afraid of "Big Bad" Generative AI? Brief Notes on the IDPA Provision Against OpenAI ChatGPT

Angela Busacca and Melchiorre Alberto Monaca

Abstract The rise of ChatGPT on the market and the rapid appearance of other similar tools proposed by companies competing with OpenAI and, more generally, by the Big Players of the web, have renewed the centrality of the debates on the use of generative AI systems, especially in consideration of the notable spread of some versions which are accessible, free-of-charge and extremely easy to use. Although progress in the technological revolution is one of the hallmarks of the current age and its positive impact is undeniable, the rise of generative AI systems appears to be characterized by an unprecedented speed, which risks multiplying socially relevant applications and overwhelming lifestyles and professional fields all too quickly without the necessary adaptation and reaction times to innovation. As is easy to see, there are multiple and complex issues connected to the use and diffusion of generative AI systems and involve not only strictly legal profiles, but also ethical and social ones, especially in relation to the possibility of massive application, with relevant consequences in different fields: in the labour market, in production processes, but also in education and communication systems and in human relations themselves. Within this complex framework of reference, the essay will focus on the dispute between Italian DPA and Open AI, with reference both to the first disputes which arose in the spring of 2023 and to the more recent disputes from January 2024, in

This essay, with the addition of notes, represents a re-elaboration and in-depth analysis of the oral presentation proposed in the "AIMED 2023" conference on 2nd December 2023; in consideration of the developments which have occurred and in particular in consideration of the approval by the European Parliament, in March 2024, of the definitive text of the so-called Artificial Intelligence Act (AI Act), some references in the text have been updated and paragraph 4, dedicated to the provisions regarding Generative AI, has been inserted. Although the results of the research come from common reflection and common discussion, paragraphs 1 and 4 are attributable to both authors; paragraphs 2 and 3 solely to Angela Busacca, who assumes full responsibility for the content and opinions expressed.

A. Busacca (✉) · M. A. Monaca
Università Mediterranea di Reggio Calabria, Reggio Calabria, Italy
e-mail: angela.busacca@unirc.it

M. A. Monaca
e-mail: melchiorre.monaca@unirc.it

D. Marino and M. Monaca (eds.), *Generative Artificial Intelligence and Fifth Industrial Revolution*, Lecture Notes in Networks and Systems 880,
https://doi.org/10.1007/978-3-031-73880-7_9

which different processing data violations and specially GDPR rule violations are hypothesized. Apart from the Italian case and the question of the safeguard of data subjects' rights, in the last paragraph some considerations of a more general nature will be presented on the AI Act (latest) text, regarding regulation of generative AI systems that use GPT and LLM algorithms.

Keywords Generative AI · Personal data protection · IDPA provision · Compliance with GDPR · Lawfulness processing activities

1 Increase in AI Generative Systems: Risk or Opportunity?

Italian statistics relating to searches carried out online in 2023 through the Google search engine [5, 23, 41][1] have highlighted how there are words and phrases relating to systems of artificial intelligence (AI), in diverse analysis sections alongside words relating to the ongoing geopolitical crises. In particular, in the "what is this?" section, at eighth place there is the term "ChatGPT", used to indicate a kind of generative AI tool produced and placed on the market by the US Company OpenAI. Since the spring of 2023, this AI tool, indeed, has been the centre of a dispute between OpenAI and the Italian Data Protection Authority (DPA), as well as being at the centre of discussions, both at a social and legal level, on the potential risks of using AI in diverse daily and professional activities.[2]

More in detail, ChatGPT is a chatbot, i.e. software designed to simulate human behaviours and, in particular, conversations through question/answer interaction, which uses a generative pre-trained transformer (GPT) algorithm. This is a linguistic generative algorithm referable to forms of Large Language Models (LLM). As a rule, LLM are AI models capable of achieving a general understanding of both language and syntax of words and processing coherent and reliable texts [48].[3]

The rise of ChatGPT on the market and the rapid appearance of other similar tools proposed by companies competing with OpenAI and, more generally, by the

[1] The statistics used relate to searches carried out during 2023 in Italy and are divided into categories according to more generic classification criteria (such as, the question: "what does it mean" or "what is this?") or more specific, identifying individual subjects or sectors (such as, politics, music or sport). The data set can be consulted at https://trends.google.com; for those relating to ChatGPT, see https://trends.google.com/trends/explore?geo=IT&q=chatgpt.

[2] Debates on the use of AI tools and in particular generative AI tools constitute the subject of a large bibliography; on the topic, see: F. Pilla, G. Cupini, M. De Santo, M. Di Maio, R. Dolce, Uno, Nessuno, ChatGPT, 2023, Ed. Ledizioni; F. Morgantini, ChatGPT—L'inizio di una nuova Era, 2023, Ed. Kennes; S. Bedessi, Intelligenza artificiale e fenomeni sociali, 2023, Ed. Maggioli; per una completa analisi sulle implicazioni della sfida tecnologica dell'AI nei diversi ambiti della vita delle comunità, cfr. L. Flordi, Etica dell'Intelligenza artificiale. Sviluppi, opportunità, sfide. 2022. Ed. Raffaello Cortina Editore.

[3] For an effective summary of the technical premises and basic functioning of the GPT algorithms, see: S. Sapienza, ChatGPT, dati e identità personale: rischi, bilanciamenti, regolazione, in Osservatorio Costituzionale AIC, 2023, p. 47 ss (ma spec. pp. 50–54, con richiami bibliografici).

Big Players of the web,[4] have renewed the centrality of the debates on the use of generative AI systems, especially in consideration of the notable spread of some versions which are accessible, free-of-charge[5] and extremely easy to use.

Precisely for its ease of use and free use, ChatGPT's presence has become immediate and pervasive in all fields of daily life, causing at the same time many doubts on possible uncontrolled uses and/or avoiding/evading legal and social rule use.

Although progress in the technological revolution is one of the hallmarks of the current age and its positive impact is undeniable, the rise of generative AI systems appears to be characterized by an unprecedented speed, which risks multiplying socially relevant applications and overwhelming lifestyles and professional fields all too quickly without the necessary adaptation and reaction times to innovation [13, 35].[6] Precisely in relation to this last point, however, it seems appropriate to highlight how the evolutionary trend of studies on generative AI has not been linear: the first experiments on software able to generate texts and simulate human conversation, indeed, date back to the 1950s and 1960s, with limited results, involving only the elaboration of short and often repetitive sentences. Up until the 1990s and even in the early years of this century, there were improvements, although the texts developed still appeared very simple, due to the training systems used. The decisive leap, which allowed us to reach current results, occurred only in the last decade, culminating with the release on the market of the first versions of the algorithms trained with pre-trained models and associated with LLMs, i.e. models of deep learning for understanding and processing human language and conversation and interaction skills, according to the question/answer scheme, with reliable text processing as well as multimedia content with very realistic images and sounds.

Taking ChatGPT as an example, we note that the first version of the GPT algorithm was released by OpenAI in 2018, therefore only four years before the release of the GPT3 version, integrated into ChatGPT and placed on the market in November 2022 and just five years before the release of GPT4, presented as a multimodal language model, which took place in March 2023.

This decisive acceleration of technological development processes is based on the two factors that make it possible to mature, strengthen and verify results in an unusually short time: on the one hand, an increase in the computational power of

[4] Among the systems currently present on the web, we can mention DALL-E and Stable Diffusion, as well as Bard, developed by Google and which will be integrated into the G-Chrome system and other systems related to it; the LLaMA projects (which involves Meta) and the Ernie Bot project (which involves Baidu) are also under development.

[5] Almost all the generative AI systems including GPT algorithms have two versions: a more complete and functional version, indicated as "pro" or "pluris", and a more basic version with free-of-charge access, but for the use of which the provider requests authorization to process different categories of personal data; on issues relating to processing activities and the use of data for training of AI systems, please refer to the following paragraph 2.

[6] A. Monreale, Rischi etico-legali dell'Intelligenza Artificiale, in DPCE on line, 2020, p. 3391 ss; A. D'Aloja, Il diritto verso il "mondo nuovo". Le sfide dell'Intelligenza Artificiale, in BioLaw Journal Rivista di Biodiritto, 2019, p. 3 ss.

computers and, on the other hand, the growing availability of data which can be collected, processed, and used for training AI systems [30].[7]

It is precisely this dual profile of the quantity and quality of data that takes on an essential value regarding the AI systems [28],[8] and in particular in the case of generative AI, since the different methods of training and self-learning of the algorithms are based on Big Data [18, 24, 32, 51],[9] which are huge quantities of data available online, and on the knowledge that can be obtained from them [1, 14, 19].[10] On this basis, it follows that any incompleteness or unreliability of the data collected and processed could have distorting effects and lead to false and, in turn, unreliable results, or even distort the entire training and learning process and, therefore, generate totally erroneous and harmful results. Thus, it is no surprise that the first public authorities to take an interest in the phenomenon and to ask for greater attention to be paid to the profiles relating to the collection and processing of data, particularly personal data, were the National Data Protection Authorities of some EU Member States and in particular the Italian and Spanish Authorities. On the findings carried out by the national authorities and in response to the need for univocal answers and a common line of intervention for the EU, the European Data Protection Board (EDPB) has set

[7] On the subject, see R. Manzotti, S. Rossi, Io&Ia. Mente, cervello e GPT, 2023, Ed. Rubettino; the authors highlight the logarithmic scale progression of the GPT algorithms over the years, underlining how the reference parameters of the algorithms have gone from a few thousand in the first version to around a thousand billion in the GPT4 version and hypothesizing, in projection, the possibility of arriving, as early as 2027, at a GPT11 version with "more than a billion parameters" which "will probably encompass all human knowledge".

[8] N. Maccabiani, Il percorso europeo verso la qualità dei dati e la sua standardizzazione nell'IA: una prospettiva legale, in BioLaw Journal Rivista di Biodiritto, 2022, p. 473 ss.

[9] The expression "Big Data" is generally used to indicate large amounts of data, coming from different collections and sources or which are the result of processing activities, which present characteristics of volume, velocity, and variety (the so called "3V") that require technologies and specific analytical methods for mining of value and knowledge. These data sets pose quite a few problems from a legal point of view both with reference to the legitimacy of the collection and processing activities, and the transfer between controllers in relation to protection of the data subject rights. Furthermore, some problems could arise also with reference to illicit uses both on the market and in professional fields. In the context of the studies on the topic, see: G. Galli, F. Sartor, L'utilizzo dei Big Data e dell'IA per una migliore qualità della regolamentazione, in Osservatorio sulle fonti, 2022, p. 341 ss; F. Faini, Big data, algoritmi e diritto, in DPCE on line, 2019, p. 1869 ss; con ottica pubblicistica, G. Mazzei, "Big Data" come beni comuni globali e principi in tema di "public utilities", in federalism.it, 2023/12, p. 162 ss; M. Tresca, "Big data", "Open data" e algoritmi, in Rivista trimestrale di diritto pubblico, 2021, p. 545 ss.

[10] On the topic, see: L. D'Avack, La rivoluzione tecnologica e la nuova era digitale: problemi etici, in U. Ruffolo (ed) Intelligenza artificiale. Il diritto, i diritti, l'etica. 2020, Ed. Giuffrè, p. 3 ss (ma spec. p. 10 ss); A. C. Amato Mangiameli, Intelligenza artificiale, big data e nuovi diritti, in Rivista italiana di informatica e diritto, 2022, p. 93 ss (https://art.torvergata.it/bitstream/2108/294409/1/98-Articolo-154-1-10-20220324.pdf); V. Ferrari, Note socio-giuridiche introduttive per una discussione su diritto, intelligenza artificiale e big data, in Sociologia del diritto, 2020, p. 9 ss.

up a dedicated task force to exchange information and to coordinate enforcement actions aimed to protect European citizens and their rights.[11]

Alongside the issues relating to data collection and processing activities, however, the more general question of the possible risks associated with the development and marketing (without limits) of generative AI systems without a specific sector regulation focuses on the draft AI Act discussed by the European Parliament. Just reading the text proposed in the early months of 2023, a significant gap emerges: generative AI was considered only marginally in the draft without the necessary attention to connected risk [21, 47].[12] Precisely to fill this gap, in June 2023 some amendments were proposed aimed at placing a series of obligations also for suppliers of basic models "used in AI systems expressly intended to generate, with different levels of autonomy, contents such as complex texts, images, audio or video". These amendments were incorporated into the AI Act and constitute a part of the text approved in the early months of 2024.[13]

As is easy to see, there are multiple and complex issues connected to the use and diffusion of generative AI systems and involve not only strictly legal profiles, but also ethical and social ones, especially in relation to the possibility of massive application, with relevant consequences in different fields: in the labour market, in production processes, but also in education and communication systems and in human relations themselves.

Within this complex framework of reference, the considerations presented here will address a specific issue, i.e., the dispute between Italian DPA and Open AI, with

[11] The press release relating to the establishment of the task force can be consulted at https://www.edpb.europa.eu/news/news/2023/edpb-resolves-dispute-transfers-meta-and-creates-task-force-chat-gpt_en.

[12] The proposal relating to regulation of AI was presented by the European Commission for the first time in April 2021, with the aim of minimizing the risks coming from the diffusion and sale of AI systems that may have a significant impact on fundamental rights and freedom of European citizens. The approach chosen by the European legislator is based on the classification of risks connected to AI systems, identifying four categories, with related duties and prohibitions for developers and suppliers and more generally, for all the subjects involved. Those four categories are: minimum risk, limited risk, high risk and very high risk; this latter category carries an absolute ban on placing and circulating it on the market. In the original text, chatbots were classified as limited risk AI tools and, consequently, subjected to reduced transparency and control duties. At that time, however, the classification did not consider the implications of GPT algorithms and LLM systems, which were already present on the market even if in the early and most incomplete versions. On the proposed law on the AI Act, see: G. Finocchiaro, La proposta di regolamento sull'intelligenza artificiale: il modello europeo bassato sulla gestione del rischio, in Il Diritto dell'informazione e dell'informatica, 2022, p. 303 ss; G. Resta, Cosa c'è di "europeo" nella Proposta di Regolamento UE sull'intelligenza artificiale?, ivi, p. 323 ss.

[13] The European Parliament expressed a vote in favour of the adoption of the AI Act in March 2024, with a very large majority: with this provision the EU "enters history" as the first legal system to regulate AI. The approved text includes a series of amendments and additions aimed specifically at generative AI systems, divided into basic models (FM—Foundation Models) and models for general purposes (General Purpose AI System—GPAI); these changes were determined precisely due to technological progress and the diffusion of LLM models and GPT algorithms, for which different levels of classification and precise transparency obligations have been indicated; for a more detailed examination of these profiles, please refer to the following paragraph 5.

reference both to the first disputes which arose in the spring of 2023 and to the more recent disputes from January 2024, in which different processing data violations and specially GDPR rule violations are hypothesized. Apart from the Italian case and the question of the safeguard of data subjects' rights, in the last paragraph some considerations of a more general nature will be presented on the AI Act (latest) text, with regard to regulation of generative AI systems that use GPT and LLM algorithms.

2 Risks of/from Unlawful Data Processing Activities: IDPA Against OpenAI

With provision no. 112, 30 March 2023 [29, 33, 48],[14] and the resulting emergency procedure, the Italian DPA imposed a temporary limitation on the processing of Italian data subjects by OpenAI, the US company that develops and manages ChatGPT software and plays the role of controller for personal data processing regarding all the activities carried out for the use, and through the use, of the above-mentioned software.

By March 2023, ChatGPT-3 had already collected more than 100 million users around the world and was the most popular generative AI with free-of-charge access. ChatGPT-3 is a machine learning-based chatbot model, with both supervised and reinforcement learning modes, capable of realizing many functions but particularly capable of generating conversations with humans, answering questions and queries and creating/writing texts on the basis of precise indications and requests.

Thanks to training carried out in the cloud (using Microsoft Azure's cloud services), ChatGPT, especially in the Chat-3 and Chat-4 versions, is able to synthesize texts based on both the processing of large amounts of personal and non-personal data, and the "absorption" of vast amounts of texts collected online and, in general, free to access and to use.[15]

As indicated on the company's website, the purpose of the OpenAI Company was to create "friendly-artificial-intelligence" (FAI) tools, easily usable by the public and able to demonstrate and promote various beneficial uses of AI. Despite this declaration, since its first presentation to the public, ChatGPT has attracted not only the attention and positive curiosity of the European public, but also some concerns and doubts due to some uses that could be dangerous, with reference to both to

[14] The text of provision can be read at https://www.garanteprivacy.it/home/docweb/-/docweb-display/docweb/9870832; first note on provision, see: A. Malaschini, ChatGPT e simili: questioni giuridiche ed implicazioni sociali, in Consulta online, 2023, p. 583; L. Megale, Il Garante della privacy contro ChatGPT: quale ruolo per le autorità pubbliche nel bilanciare sostegno all'innovazione e tutela dei diritti? In Giornale di diritto amministrativo, 2023, p. 403; S. Sapienza, ChatGPT, dati e identità personale: rischi, bilanciamenti, regolazione in Osservatorio costituzionale, 2023, p. 47.

[15] Users can activate an account to use the demo or basic/free versions or the more complete ones which require a paid subscription and the possibility, of course, to enjoy better performances and more accurate and complete results/texts.

processing of personal data but, above all, possible uses for illicit purposes or the dissemination of fake content.[16]

Moreover, it should be underlined that, alongside the fake contents that can be knowingly created and used for illicit purpose, i.e. to harm fundamental rights of people or cause personal damage or alter public opinion, ChatGPT has also reportedly created false content generated by errors in data processing or by unreliable data. On several occasions, users have received clearly erroneous answers, with news and facts incorrectly indicated or even with the result of an invention with no evidence. Hence, the spotlight is now on how the question regarding data collection and data processing activities for AI training plays a central role. Conscious illicit uses undoubtedly constitute an alteration compared to normal uses aimed at improvement of human activities and making knowledge even more accessible. Furthermore, the possibility of wrong or misleading output could depend on quality and quantity of training data sets and, moreover, on the quality and reliability of data collected online, on other sources checked and on the modalities of training of AI tools.

The critical point in the Italian DPA provision regards the modalities of collecting and using personal data of EU users and the unclear clauses regarding data subject information, as well as the fulfilment of those duties of controller connected to the processing activities in training AI tools. The Italian DPA limitation measure on data processing in provision n. 112/23 was adopted pursuant to Art. 58, par 2 lett. f GDPR [8, 10],[17] following a data breach (loss of conversation/data for 200,000 users) and the start of an inquiry,[18] which highlighted the presence of at least three critical elements connected to the fulfilment of the duties/obligations imposed by

[16] While, indeed, the various functions of ChatGPT could be useful for the dissemination/improving of knowledge and for accessing contents, as well as for the level of complexity in the (simulation of) conversation and in answering questions, already in the first months of use, some risks have emerged, and, in particular, risks associated with illicit uses or uses that, in any case, could be prejudicial to persons. In relation to this last aspect, just consider the possibility of using ChatGPT for carrying out work activities or for drafting texts for school or training activities: these are very realistic hypotheses, as evidenced by the increasingly widespread and incisive adoption in the field of recruiting and human resource management. Indeed, in some States of the US, bans have been imposed on the drafting of school papers, essays and university publications. In relation to possible illicit uses, it is important to take into consideration the processing of fake contents to be introduced into information circuits or the creation of false profiles or models for phishing or other offenses aimed at the theft of personal credentials.

[17] Article 58 GDPR focuses on inquiry powers of national supervisory authorities; in particular, paragraph 2 states that each supervisory authority shall have some corrective powers, among which, as in letter (f), to "impose a temporary or definitive limitation including a ban on processing". On the topic, see: A. Caselli, Articolo 58. Poteri, in G. M. Riccio, G. Scorza, E. Belisario (editors) GDPR e normativa provacy, WKI, Milano, 2022, p. 596; F. Cardarelli, Articolo 58. Poteri, in R. D'Orazio, G. Finocchiaro, O. Pollicino, G. Resta (editors) Codice della Privacy e Data Protection, Giuffrè, Milano, 2021, p. 741.

[18] The inquiry was closed in January 2024 with notification of breach of GDPR rules by the Italian DPA to OpenAI; press release note can be read at https://www.garanteprivacy.it/web/guest/home/docweb/-/docweb-display/docweb/9978020; the notification is still a reserved document, but it is easy to image that it contains the same contestation of provision n. 112 on the breaches of data protection law: just reading the breach rules it appears clear that they are the same as provision n. 112 and provision n. 114 of 2023.

the GDPR on the controller relating to collection and processing of personal data. In response to the provision, OpenAI suspended access to the ChatGPT software for users in Italy, but at the same time indicated the will to find a solution to improve safeguard measures and transparency in data collecting and processing activities and to solve all other critical points. In March 2023, OpenAI and the Italian DPA held meetings, also via videoconference, to collaborate in finding a solution for the safe and licit use of ChatGPT from Italy.[19] During those meetings, the Italian watchdog requested more transparency and stronger measures to protect users/data subjects' rights, especially regarding underage users. On their part, OpenAI affirmed that collection and processing activities on users' personal data were carried out to improve services and not for commercial or marketing purpose, thus perfectly respecting GDPR rules, but at the same time affirming the will to respond to the Italian DPA contestation improving safeguard measures [44].[20]

At the end of diverse meetings, the Italian watchdog released another provision, n. 114/2023 of 11th April, which suspended the previous limitation of provision n. 112/2023 and ordered OpenAI to carry out some safeguarding measures and transparency improvements. At the same time, the Italian DPA, going on with the inquiry on GPDR violations, asked OpenAI for a report on the measures undertaken, by the end of May 2023. Following the indication of provision n. 114/2023, OpenAI has made ChatGPT-3 (and ChatGPT-4) accessible from Italy.

In the following months, the use of all ChatGPT software increased, as well as the use of other generative AI software, but the Italian DPA inquiry continued until January 2024, when the Italian watchdog notified OpenAI of another violation of data protection law: based on fact-finding activity and following the temporary ban on processing of provision n. 112/2023, indeed, the Italian DPA affirmed the existence of several GDPR breaches.

With this notification, the procedure for GDPR infringement against OpenAI is now open (as at end of January 2024); the company has time (30 days from notification) to present counterclaims concerning the alleged breaches that will be analysed by the Italian DPA also in the light of the work in progress of the ad-hoc task force set up by EDPB to rule on the final determination of the case.

---

[19] During the access block imposed by OpenAI, various modalities were identified to circumvent the geographical block referring to connections from Italy; for example, it was possible to connect to a virtual private network (VPN) which relocated the connection to a country where there were no restrictions and by these VPNs get access to ChatGPT.

[20] On this topic, see: E. Prosperetti, La disciplina contrattuale dell'uso dell'IA generativa. Una lettura delle nuove condizioni d'uso dei servizi di OpenAI, in Diritto dell'Internet, 2024, p. 211. The author highlights how the changes made by OpenAI can be traced back to the ongoing disputes with the National DPA (specially the Italian one) but not directly to the Digital Service Act (EU Regulation 2022/2065) which OpenAI "seems not to consider applicable to its services": precisely on this last point, however, it is legitimate to raise doubts since the activities could be classified as hosting services since they are "stored and processed, at the user's request, information provided by the user" (p. 211: the author highlights reference to the definition offered by the DSA in Art. 3, letter (g) n. iii) (?).

3 What IDPA Asked—What OpenAI Answered

Provision n. 112/2023 and the subsequent GDPR breach notification of January 2024 essentially focused on three points: lack of information given to the data subjects; use of personal data for the purpose of training the ChatGPT algorithms without suitable lawfulness and suitable guarantees and inaccurate processing activities; absence of checks and controls for compliance with the minimum age required for access to the service and the possible risks for minor users. In consideration of the critical issues found, the Italian DPA identified violations relating to Articles 5, 6, 8, 13 and 25 of the GDPR.[21]

By analysing the Italian watchdog's complaints in more detail and making a comparison with the answers offered by OpenAI, it is possible to offer some considerations on each of the critical points highlighted and evaluate whether the current situation can actually be said to be satisfactory regarding respect for EU citizen/user rights.

(a) *Information to be provided where personal data are collected from the data subjects*

The first point concerns the lack of complete information to be provided where personal data are collected from the data subjects, pursuant to Art. 13 GDPR. As highlighted by doctrine [20, 27, 50] [22] and reiterated by the courts and by some EU Member State National DPA [49],[23] also with a systemic reading of the rules on data protection, the information-to-data subjects performs the fundamental task of making the data subjects truly aware of the processing activities that will be carried out on their personal data. In this way, they can give an informed and valid consent, and, therefore, the lack of this information renders the full data collection and processing flawed as it is carried out without the data subjects being aware of modalities and purposes of processing activities. The omission and incompleteness of the information are considered unlawful conduct and make the controller, who is the subject implicated in Art. 13 GDPR, liable to sanctions pursuant to Art. 83 GDPR.

[21] The indicated articles concerned: "principles" of processing activities (Art. 5), "lawfulness of processing" (Art. 6), "conditions applicable to child's consent in relation to information society services" (Art. 8), "information to be provided where personal data are collected from the data subject" (Art. 13) and "data protection by design and by default" (Art. 25).

[22] L. Liguori, Informazione e accesso ai dati personali, in R. D'Orazio, G. Finocchiaro, P. Pollicino, G. Resta (eds.), Codice della Privacy e Data Protection, Milano, 2021, p. 287; P. Todaro, Art. 13—Informazioni da fornire qualora i dati personali siano raccolti presso l'interessato, in G. M. Riccio, G. Scorza, E. Belisario (eds.), GDPR e normativa Privacy, II ed., Milano, 2022, p. 175; G. Finocchiaro, Il nuovo Regolamento europeo sulla privacy e sulla protezione dei dati personali, Torino, 2017 (especially, Il quadro d'insieme sul Regolamento europeo sulla protezione dei dati personali, p. 1–26).

[23] V. Silvetti (ed), I provvedimenti in materia di data protection dei Garanti Europei, Milano, 2021 (especially, § 6 Violazione obbligo di informativa, p. 45, § 7 Omesso obbligo di informativa, p. 49, §32 Informativa e diritto di opposizione, p. 211).

After meeting with the Italian DPA, OpenAI updated both the terms of use and the privacy policy, the links of which are present on the home page of the website at https://chatgpt.com/auth/login. Regarding the privacy policy, currently there are two versions of the information: one updated on 15th Feb 2024 for users living in the European Economic Area (EEA), in Switzerland and in the UK; and another, updated to 31st Jan 2024 for users living elsewhere.

There follows a synthesis of what can be read on the website, in the most general terms regarding the current information given to data subjects:

- indicates the practices adopted for collection of personal data from users/data subjects or from users of the OpenAI website, applications and services; however, data that are processed on behalf of customers who use commercial offers, such as APIs, are excluded, since the use of such data is regulated by agreements with the customers themselves;
- indicates that personal data is collected and used for training and development purposes of the linguistic models that power ChatGPT and other services;
- indicates the source of collected data that for ChatGPT are information which has become public online, information provided by third parties and information provided by users or human trainers. Regarding the different categories of information and data that could be object of collection, these include: data provided by the user, data present in user content, such as uploaded files or feedback, information present in communications with users, information present in social media pages and also information provided indirectly by users, through their participation in online activities, events or surveys;
- specifies, also in type font that have particular graphic emphasis, that personal data are collected and used only to improve the functionality of the language models, both in the understanding and in response phase, also highlighting that no processing is envisaged for profiling purposes or for commercial or marketing purposes;
- specifies that each processing activity is carried out in compliance with the legislation on privacy and data protection and that the rights of the data subjects are guaranteed, including the right of opposition; moreover, there are also a series of links and contact addresses both to obtain further information and to assert these rights or complain about a violation.

Already at a first glance, this information appears to be sufficiently clear and comprehensible, it uses simple terms and a colloquial and direct tone which makes it easy to read, and also uses bold graphics for those phrases that deserve particular attention, although in several points it requires passages from one section to another through links and may sometimes create some confusion: on paper, Article 13 of the GDPR is respected and right to be informed of data subjects is guaranteed. On paper.

However, the critical point is the real usefulness of the information which is structured for the hypotheses of data collection and processing activities for purposes of training AI, and, particularly, generative AI [38, 43].[24] It conveys a series of notions to users which are either provided in a technical way, and therefore require a long text which may be difficult to read, or are overly simplified, so much so that they lose their real informative value. Doubts, already expressed in doctrine [6, 22, 26, 34] [25] on the need for an update of the GDPR, arise again, especially in relation to the processing carried out for issues connected to AI tools. It has already been highlighted, through surveys and practical studies, that many users do not pay attention to privacy policies

[24] On the topic, see: E. Prosperetti, ChatGPT ed altri tool di AI. Condizioni di utilizzo, questioni di proprietà intellettuale e privacy, in Diritto dell'Internet, 2023, p. 245 (but especially § 6: l'uso dei dati personali nei servizi AI esaminati); for a more general and complete approach regard using data for AI training, see: M. G. Peluso, Intelligenza artificiale e tutela dei dati, Milano, 2023.

[25] The dispute over inadequacy of some parts of the GDPR regarding this topic mainly concerns Art. 22 on "automated individual decision-making, including profiling": the data subject has the right not to be subjected to decisions based solely on such processing, if such decisions produce legal effects concerning him/her or significantly affects him/her (paragraph 1); furthermore, in the case of decisions that are necessary "for entering into, or performance of, a contract between the data subject and a data controller" or "is based on the data subject's explicit consent", the controller has the duty to implement measures that are adequate to protect the rights and freedoms of the data subject and this latter has the right to request human intervention on the part of controller, to express his/her point of view and to contest the decision (paragraphs 2 and 3). A first question, which arose among scholars already in the first GDPR analyses, concerns the real scope of Article 22 in relation to the adverb "solely" which could exclude from the application of this rule all those processing activities in which minimal human intervention can be recognised, thus determining a protection vulnerability in the case of processing activities where automation is almost total. Nor, on the other hand, would other methods of protection be discernible, given that the so-called right to algorithmic explanation not only does not appear to be expressly contemplated in Art. 15, among the rights of the data subject, but even if it were, it would not constitute a strong protection, since the explanation might be incomprehensible as it is extremely technical. On this topic, see: F. Lagioia, G. Sartor, A. Simoncini, Art. 22—Processo decisione automatizzato relativo alle persone fisiche, compresa la profilazione, in R. D'Orazio, G. Finocchiaro, O. Pollicino, G. Resta, Codice della Privacy e Data Protection, cit., p. 378; R. Messinetti, Comunicare nell'infosfera. La vulnerabilità della persona digitale, in federalism.it, 2021/18 (https://iris.uniroma1.it/retrieve/e383532d-2cb8-15e8-e053-a505fe0a3de9/Messinetti_comunicare_2021.pdf); G. Finocchiaro, Riflessioni su Intelligenza artificiale e tutela dei dati personali, in U. Ruffolo (ed), Intelligenza artificiale. Il diritto, i diritti, l'etica, Milano, 2020, p. 243; L. Bifulco, Intelligenza artificiale, Internet ed ordine spontaneo, in F. Pizzatti (ed) Intelligenza artificiale, protezione dei dati e regolazione, Torino, 2018, p. 393.

On these premises and considering the methods of creating training datasets for algorithms and machine learning of generative AI systems, it can be stated that we are in the presence of automated processing for which none of the aforementioned rights are recognized to the data subject; nor could it be otherwise, considering the amount of data used for training and the processing that the data undergoes for this purpose. In extreme summary, it can be stated that the perspective of the GDPR appears difficult to be reconciled with the activities related to the development of AI, since many provisions of the GDPR are based on approaches that consider the individual dimension of the controller–data subject relationship and which finds points of criticality on the occasions of more complex relationships, when through successive passages of data sets or through the intervention of automatic methods, other subjects come into consideration or the same individual person as data subject is lost in a larger and more indistinct group.

In this sense, it seems clear that effective protection of the rights of data subjects with regard to data processing activities aimed at training algorithms for AI tools must call into question any intervention by the European legislator, who acts by enriching the GDPR with further provisions capable of responding to the needs emerging from the spread of AI systems.

particularly regarding the information pursuant to Art. 13 GDPR. Indeed, often, users agree to all proposed data process requests without having even a basic knowledge of the different characteristics of the processing activities for which consent is given. These issues gain renewed importance when it comes to AI tools, since processing activities, especially those aimed at training algorithms, present a particular degree of risk to user/data subject rights.

The question of information is reflected in the more general question of updating the GDPR and, in parallel, in the provision of a series of obligations on the controller, who is also a supplier of the AI software. The rapid evolution and extreme technicality of the issues inherent to the training methods and operation of AI systems, in fact, would require a far more complete digital literacy on the part of users for the purposes of understanding technical information and, consequently, of a truly informed consent: however, this situation is decidedly utopian.

The protection of users/data subjects must necessarily be based on compliance with the rules of transparency and risk limitation by the developers and suppliers of the AI tools: looking only at the rules of the GDPR does not appear sufficient and, therefore, not just an update, but also that such an update takes place considering the AI Act, is desirable.

(b) *Lawfulness and inaccurate outcome of processing activities*

In provision n. 112/2023, the Italian DPA also noted "the lack of a suitable lawfulness for the processing of personal data in relation to the purposes of training the algorithms underlying the functioning of ChatGPT": in the version online in March 2023, indeed, this lawfulness was related to "the performance of a contract to which the data subject is party", as referred to Art. 6, n. 1, letter b GPPR.

The acceptance of the "terms and conditions of use", for the free versions, and the signature of the "subscription conditions", for the paid versions, place the data subject in the role of party to a contract and this could justify the processing for data strictly connected to requested services but certainly not for data collection aimed at training the algorithms, since at the time of the conclusion of the contract the algorithm underlying the functioning of ChatGPT has already been trained and ready to use.

Reference to the performance of a contract which the data subject is party to does not appear suitable for the hypothesis in question, since the processing of data is aimed at further training of the algorithm, therefore, not for purposes directly connected to the performance of an actual contract. Nor can the explanation offered by OpenAI, i.e. the implementation of services, be considered valid since it could, at most, refer to the updating of the service, therefore always with a future perspective and only possibly regarding the contract in act.

On this point, the Italian watchdog has made a request to modify the reference for lawfulness, just indicating the different hypotheses of the data subject's consent (Art. 6, n. 1, letter a) and of the legitimate interest pursued by the controller (Art. 6, n. 1, letter f), thus amending reference to the performance of the contract which the data subject is party to (Art. 6, letter b) [7, 42, 47].[26] This request was only partially accepted, since, in the privacy policies currently online, reference to the performance of the contract which the data subject is party to is still present, although it is alongside the different hypotheses of the legitimate interest of the controller, of compliance with a legal obligation which the controller is subject to, and only marginally, the consent of the data subject.

In particular, reading the current privacy policies on the website, OpenAI bases personal data processing on:

- the performance of the contract which the data subject is party to, for the purposes of supplying and maintaining services, as well as communications with users, for information on services and events;
- the legitimate interest of the controller: to improve and develop the services offered, for new features and to conduct research;
- compliance with a legal obligation which the controller is subject to: to prevent fraud, criminal activity or abuse and to protect security of the services, as well as to protect the rights, privacy, safety and property of users, affiliates or third parties;
- the consent of data subjects: when the processing is aimed at a specific purpose that is expressly communicated, such as, the sending of communications for marketing purposes.

In relation to the lawfulness of processing, it seems appropriate to highlight that OpenAI provides continuous updates; at first, in fact, the consent of the data subjects was not considered and in provision n. 114/2023 the Italian DPA highlighted how OpenAI's response had only been partial, asking them to make further changes and to eliminate references to the performance of the contract which the data subject is party to.

Analysing the privacy policies updated to 15 Feb. 2024 and currently online, both the persistence of the reference to the performance of the contract and the absolute marginality of the data subject's consent emerges, while reference to compliance with a legal obligation which the controller is subject to appears decidedly broad with, only subordinately, reference to the controller's legitimate interest.

Some considerations can be made on this point: the choice of conditions for lawfulness other than consent, really, is not surprising since, precisely in consideration of the extreme technicality of the issues, it appears difficult to configure a valid consent pursuant to Article 7 of the GDPR[27] that presents all the elements required

[26] Generally, on Art. 6 GDPR, see: D. Poletti, Art. 6—Liceità del trattamento, in R. D'Orazio, G. Finocchiaro, O. Pollicino, G. Resta, Codice della Privacy e Data Protection, cit., p. 208; F. Resta, Art. 6—Liceità del trattamento in G. M. Riccio, G. Scorza, E. Belisario (eds), GDPR e normativa privacy, cit., p. 78; F. Bravo, Il consenso e le altre condizioni di liceità del trattamento dei dati personali, in G. Finocchiaro (ed), Il nuovo regolamento europeo sulla privacy e sulla protezione dei dati personali, cit., p. 101.

[27] Art. 7 states that in any case of processing activities based on consent, the controller "shall be able to demonstrate that the data subject has consented to processing of his or her personal data"; moreover from the second paragraph it emerges that "if the data subject's consent is given in the

by the definition of Art. 4, n. 11 GDPR, i.e. to be free, specific, unequivocal and informed [15].[28] Moreover, reference to the performance of the contract which the data subject is party to or to compliance with a legal obligation which the controller is subject to, allows to shift the system equilibrium to the side of the "strong" part of the contractual relationship and to overcome the uncertainties coming from the validity of the consent, especially in the case of processing activities subsequent to, and further than, those envisaged at the time of obtaining consent.

Some considerations can also be proposed on this point and, once again, that sort of obsolescence that characterizes the GDPR emerges in reference to issues inherent to AI tools and, on the topic of consent, recalls a series of doubts which had already arisen in doctrine with reference to automated processing pursuant to Art. 22 GDPR.

Regarding the training of algorithms, however, the scope of validity of the consent would be even more limited in consideration of the risks related to the development of AI: the data subject, indeed, would be asked to give consent to processing activities in relation to which not even the controller could have full awareness of the consequences. On this basis, there would be two possibilities: either to admit consent that is not really fully informed, or to admit consent becoming rapidly obsolete. In the latter case, a situation would be created that required the need for a continuous renewal of the consent to allow the use of the personal data in subsequent training.

However, still in reference to a data subject's consent, an additional profile of the "material" scope of consent emerges since the processing activities aimed at training project integrating personal data into the so-called "knowledge value chain", through which data mining activities allow us to move from raw data to information and to knowledge. As raw personal data and information have been processed and become "knowledge", they are no longer considered individually but only in reference to the overall value of the knowledge as a new IT good, which is valuable from both an economical and a legal point of view. From this consideration, the doubt arises whether the individual data subject consent still has relevance or, in just considering the information and the knowledge as the result of controller activities and as new informational goods, processing activities carried out on them does not require further consent and/or authorization from data subject, i.e. the natural person to which the raw data belong.

Taking this point of view, the choice of the controller's legitimate interest would seem to solve all the problems relating to lawfulness; however, even with reference to this hypothesis, critical aspects emerge, since this criterion could lend itself to exploitation and abusive use of data subject's rights, as referred in Art. 6, n. 1 lett. f. In this regard, reference can be made to the juridical precedent of the "Clear-view"

context of a written declaration which also concerns other matters, the request for consent shall be presented in a manner which is clearly distinguishable from the other matters, in an intelligible and easily accessible form, using clear and plain language".

[28] On Art. 4, n. 11, see: A. De Franceschi, Art. 4—Definizioni, in R. D'Orazio, G. Finocchiaro, O. Pollicino, G. Resta, Codice della Privacy e Data Protection, cit., p. 156 (especially, paragraph 12: Consent, p. 166).

case [25, 31, 40] [29]: an excessively extensive interpretation of a controller's legitimate interest would, in fact, end up bringing interests of mere commercial nature back to lawfulness which cannot go beyond the limits set to protect the rights and freedoms of the data subjects. (*Not clear*-I know, it was in the last version, but on re-reading it's not so clear!)

---

[29] With ordinance n. 50, 10 February 2022, the Italian DPA imposed a fine of 20 million euros on the US-based company Clearview AI Inc. for privacy law breach, contesting multiple violations including the illicit collection of personal data as well as the processing of personal data, including profiling, in the absence of a valid legal basis, using a facial recognition AI software.

By techniques of web scraping, i.e. massive collection of personal data from different online sources, and using an AI software developed by itself, the Clearview company had created a database containing over 10 billion images of physical subjects, effectively carrying out a biometric monitoring and profiling activity which could be accessed through a search engine to create "one-to-many" searches: once the search criteria were entered, the system started comparing the thousands of biometric templates created and indexed and returned the outcomes based precisely on the biometric profiling activities, therefore on the processing of particular categories of data (referred to in Art. 9 GDPR). To create the database, the Clearview company declared that it had used images of people living all over the world, therefore also in Europe and Italy, drawing from "public web sources such as information sites, social media and online videos" and had compared its activity to that of indexing carried out by Google Search, implicitly considering the legitimate interest of the controller as the legal basis for processing activities. The Italian DPA, however, did not consider such conduct to be lawful, indeed it identified a multiplicity of breaches of privacy law: the database, in fact, contains thousands of profiles based on biometric data extracted from the images and subsequently enriched by other related information, such as title and geolocation of the photo and publication web page. Furthermore, this database is constantly updated and enriched with new information, therefore it is not limited to the static archiving of personal and biometric data, but rather constitutes a sort of dynamic archive, through which comparison and control activities of data subjects are carried out. The Italian DPA contested that collection and processing activities are carried out in the absence of a legal basis: the use of web scraping techniques, indeed, does not require direct contact with the data subjects, so as to be able to provide information pursuant to Art. 13 GDPR or acquire consent pursuant to Art. 7 GDPR, but provides for the massive collection of images and biometric and personal data uploaded and present online for other and different purposes; the consent or the different legal basis that make the original posting/uploading on the net lawful cannot also be assumed as a basis for the collection and activities carried out by Clearview, nor can it be considered that the online presence of the images makes themselves freely usable, depriving the data subjects of any right or protection: as reiterated by the Italian DPA, web scraping activity itself is considered illicit and prohibited by various providers and managers of platforms and social networks. At the end of the inquiry, the Italian DPA imposed not only a fine of 20 million euros, but, regarding the data of natural persons located in Italy, ordered: a ban on continuing the collection and archiving of images; the interruption of any further activity on the data already collected; the deletion of data already collected and processed; the identification of a representative in Italy, who can act as an interlocutor with the data subjects for guaranteeing their rights; communication of the initiatives undertaken to implement the ordinance.

The Clearview case undoubtedly represents an important point in the affirmation of the rights of data subjects against the massive data collection activities on the web and draws attention to the need for more efficient guaranteed tools with regard to technological developments which can entail ever greater risks for those interested. On the Clearview case, see: J. Piemonte, V. Papakostantinou, Il caso Clearview AI: uno stress test per il Regolamento generale per la protezione dei dati e la proposta di Regolamento sull'intelligenza artificiale in relazione alle nuove sfide poste dal riconoscimento facciale, in Cibespazio e Diritto, 2023, p. 43; B. Marchetti, L. Parona, la regolazione dell'Intelligenza artificiale: Stati Uniti e Unione Europea alla ricerca di un possibile equilibrio, in DPCE on line, 2022, p. 237; D. Iacovelli, M. Fontana, Nuove sfide della tecnologia e gestione dei rischi nella proposta di

It cannot be excluded that, in the continuous updating of the privacy policies that OpenAI is carrying out to respond to the complaints of the various national DPAs, further changes will be made also to confirm the lawfulness of processing personal data. However, the most critical point remains the processing aimed at training the algorithm; on this topic it would be appropriate to re-read the rules on data protection in light of the AI Act, but also in the hope for concrete and rapid interventions by the task force established by the EDPB.

Closely linked to the lawfulness of personal data collection and use, there is a further point highlighted by the Italian DPA: inaccuracy of the processing activities and, consequently, inaccuracy of the proposed outcomes. Based on various fact-checks carried out, the Italian watchdog was able to verify that the information provided by ChatGPT does not correspond to real situations, giving erroneous and unreliable outcomes and/or answers. In this regard, OpenAI reiterated its commitment to improving the functionality and security of its products and services, whilst being aware of the limits of laboratory activities in the field of AI.[30] The question of outcome reliability, moreover, is directly connected to two further questions: data quality and reliability of the data collection and processing criteria in the so-called knowledge value chain which goes from raw data to knowledge capable of orienting the most efficient and informed choices. The data mining process that leads to knowledge, in fact, can lead to distorted outcomes both by erroneous or false or corrupt data, and by processing modalities that determine erroneous combinations and fake associations between the data, leading to outcomes not responding to real data; those false outcomes represent a serious risk, as they can distort personal opinions and beliefs and moreover contribute to spreading misinformation and lack of knowledge.

(c) *Verification of the age of users*

A further point of dispute by the Italian DPA relates to protection of underage users and lack of methods for ascertaining the actual age of users, in order to avoid circumvention of rules on the minimum age for access to services. With regard to this last profile, indeed, the general rule provides that in the case of underage users, the double requirement of parental consent and the minimum age of 13 must be satisfied: this double limit is not only linked to a generic duty of supervision of parents over minors but provides for a minimum age threshold precisely to avoid exposing any younger users to serious risk [3, 45].[31]

---

regolamento europeo sull'intelligenza artificiale: set di training, algoritmi e qualificazione dei dati. Profili critici, in Il diritto dell'economia, 2022/3, p. 106.

[30] On the ChatGPT web site it is possible to read that Open AI "*also recognizes that, like any technology, these tools carry real risks, so it works to ensure that security is built into the system at all levels (…) We work hard to prevent predictable risks before deployment, however, there is a limit to what we can learn in a lab. Despite extensive research and testing, we cannot predict all the beneficial ways people will use our technology, nor all the ways they will abuse it. That's why we believe that learning from real-world use is a critical component to building and releasing increasingly secure AI systems over time*".

[31] To verify the age of users of online services, various methods and procedures can be used; most of them are based on document checks or personal security question checks or even on access

This issue is connected to the more general theme of the protection of underage users with regard to ICT tools both from the point of view of the protection of personal data and from the point of view of the conscious and responsible use of devices, software and applications [2, 11, 12, 39].[32] The rapid spread of apparently easy-to-use and free software and applications that allow to modify images and create audio and video content through AI tools, has, in fact, multiplied the dangers related to use by minors and made these dangers even more evident, particularly, in use by pre-adolescents. The risks come not only from malicious use, deliberately aimed at causing offenses and damage, but also from any superficial uses that could be dangerous and cause damage, or even from the conveyance of artificial and distorted models capable of affecting the existential and emotional sphere of minors, causing or increasing a sense of frustration and inadequacy, with negative effects on self-perception and personality development. The provision of a minimum age for users of the services, therefore, appears necessary precisely to guarantee a degree of maturation and ability to understand the potential of the AI tools which should allow the minor, presumably under the control of parents, to correctly use ChatGPT services.

On this point, following the Italian DPA provision of 30 March, a declaration appeared on the OpenAI open page which places policies for the protection of minors as a priority, indicating some changes which had already been developed in the ChatGPT-4 version and some actions being completed: among the latter also new "verification options" of the actual age of the users. Of particular interest is the fight against forms of abuse and the online circulation of child pornography material; this is carried out, on the one hand, by minimizing the potential for generating content that is harmful to children and, on the other hand, by connection with the National Centre for Missing and Exploited Children, which any attempt to create and upload child pornography materials in image form are reported to.

---

to official registers, but in most cases, these are merely formal checks. The declared purpose is to protect minors from access to online network or contents that may be dangerous both for the possible implementation of harmful conduct (think of applications for image manipulation) and for the spread of data that may be stolen and used for illicit purposes. The most critical point is precisely the effectiveness of these systems, since too often minors manage, alone or with the complicity of adults (sometimes even their parents) to overcome the limits imposed by the regulations or the conditions of use of the services and to access online services. On the topic, especially see: C. Arnone, L'uso dei servizi di intelligenza artificiale da parte dei minori: solo una questione di "age verification"?, in Diritto dell'Internet, 2024, p. 13; E. Raffiotta, M. Baroni, Intelligenza artificiale, strumenti di identificazione e tutela della identità, in BioLaw Journal, 2022, p. 165 (https://teseo.unitn.it/biolaw/article/view/2245/2226).

[32] The protection of minors in the Internet ecosystem represents one of the most studied topics; within a very large bibliography, some recent contributions are highlighted: G. Citarella, A. Venchiarutti, Diritti della personalità del minore e tutele nella rete Internet, in M. D'Auria (ed), I problemi dell'informazione nel diritto civile, oggi. Studi in onore di Vincenzo Cuffaro, Roma, 2022, p. 527; G. Pesci, Prospettive e recenti sviluppi della tutela dei minori on line, in Ciberspazio e diritto, 2022, p. 311; D. D'Alfieri, Internet: quando la "rete" cattura i minori, in Rivista Italiana di informatica e diritto, 2022, p. 53; E. Andreola, Minori e incapaci in Internet, Napoli, 2019.

Regarding protection of underage users, the regulatory references, in the EU legal framework, are GDPR on data protection ([4, 9, 50],[33] and the Digital Services Act, with reference to the dissemination of content and the limitation of suggestions based on profiling. Alongside these regulations, a separate consideration must be made for the rules to combat situations of exploitation for sexual purposes and the conveyance of child pornography content: in this context, indeed, it is necessary to identify the hypotheses in which the creation of artificial content can take on criminal relevance.

Precisely in consideration of the particular situation of minors and the vulnerabilities related to personality development and growth, duties and obligations on providers and content suppliers must be more stringent. With specific reference to the needs of data circulation, training of algorithms and conveyance of contents created with AI, a dynamic circuit is set in motion to which the existing legislation is not always able to offer adequate responses in terms of guarantee and protection of users.

Age verification is confirmed as an important protection tool, considering the large amount of data, including biometric data, that can be collected, increasingly used, and "captured" by applications that use photos and video live recording or by access to content galleries stored on devices.

The continuous updating work implemented by OpenAI should offer better guarantees also with reference to underage users. However, there has been no lack of criticism highlighted especially by those who affirm how the assessment methods can be easily circumvented and, above all, how the data processing and data mining activities for training purposes are in any case opaque; this last consideration, however, re-proposes the problem of quality and quantity of data collected in relation to their reliability.

[33] The GDPR provides for the protection of minors' personal data mainly in Art. 8, identifying the "conditions applicable to the consent of minors in relation to information society services". With reference to the minimum age for consent, Art. 8 indicates the threshold of 16 years and requires, for minors who are younger than that, consent or authorization from the parents or persons caring for the minor; considering this generic provision, however, the last sentence of paragraph 1 allows EU Member States to also indicate lower ages, as long as they are not under 13 years. This consent mechanism, however, has not always proven to be effective, since too often the consent of minors is not at all informed and aware, and by exposing minors to risks and dangers, it manifests all the limitations due to the lack of serious "media education". On this topic, see: E. Battelli, Privacy e minori: l'inadeguatezza del consenso digitale, in M. D'Auria (ed), I problemi dell'informazione nel diritto civile, oggi. Studi in onore di Vincenzo Cuffaro, Roma, 2022, p. 543; more generally, on Article 8 GDPR, see: C. M. Cascione, Art. 8—Condizioni applicabili al consenso dei minori in relazione ai servizi della società dell'informazione, in R. D'Orazio, G. Finocchiaro, O. Pollicino, G. Resta, Codice della Privacy e Data Protection, cit., p. 229; G. Spoto, Disciplina del consenso e tutela del minore, in S. Sica, G. D'Antonio, G. M. Riccio (eds), La nuova disciplina europea della privacy, Milano, 2016, p. 211.

4 Notification to ChatGPT on Breach of Privacy Law and Opening of Investigation into "Sora". Italian DPA Versus OpenAI: And the Story Continues...[34]

As already mentioned, the Italian DPA ban on personal data processing activities imposed on OpenAI and the consequent interruption of accessing the ChatGPT service platform were removed as early as April 2023, following meetings between the company and the Italian watchdog also in consideration of the reassurances offered by OpenAI on the adaptation of privacy policies and modalities for both verifying users' age and protecting minors.

In the following months, despite the changes actually made by OpenAI with the updating of terms and conditions of use, as well as privacy policies, doubts about the risks and dangers related to the use of generative AI have not diminished, indeed, the Italian watchdog continued its investigative activity, closing the inquiry with the notification to OpenAI, in January 2024, of the breach of privacy law regarding the web services offered by ChatGPT. This notification formally started the procedure for the assessment and possible sanctions. In the coming months, the issue will be defined also considering the decisions of the specialized EDPB task force, which has already completed part of the work but has not yet made the results known.

The story relating to ChatGPT, however, is not the only issue that focuses attention on personal data processing and data subject protection in the face of the development of generative AI systems. A few weeks after the notification of the dispute, on 8 March 2024, the Italian DPA notified OpenAI of a further provision relating to the request for information regarding "Sora", a new AI model that can create highly detailed video content with "realistic and imaginative" dynamic scenes, based on few textual instructions.

In particular, the information requested concerns primarily the availability and accessibility of the service for individuals located in the EU, and in particular in Italy, and OpenAI will have to clarify whether the service is already accessible or whether it will be and in which mode. Considering its probable diffusion also in the EU, the Italian DPA asked to indicate whether the methods of information and the legal bases for processing personal data will comply with the rules of the GDPR. With reference to personal data in particular, regarding the functioning algorithm of "Sora", the question of the training methods and the selection and collection of data used for this purpose are called into question: which kind of personal data will be processed, only personal data pursuant to Art. 4 or also particular categories of data pursuant to Art. 9 and moreover which kind of source will be used to find and collect data?

In any case, the central issue remains the compilation of the training datasets intended for the generative AI algorithms. "Sora" offers the creation of videos with

---

[34] This paragraph was added in April 2024, after the Italian DPA notification of breach of privacy law and the Italian DPA information request regarding Generative AI "Sora"; it replaces the last part of the oral presentation, dedicated to supposition and ideas on the dispute between the Italian DPA and OpenAI.

realistic content and not only with synthetic or "gamified" images, thus it determines the need to draw on an enormous amount of data, especially biometric data that allows to recreate facial and body movements. A possible solution could be the use of so-called synthetic data, in relation to which, however, the areas of use still remain marginal and the costs very high, without also considering the total absence of any regulation.

Regulation itself becomes the centre of gravity of the issue, especially following the approval of the AI Act which, while representing the first law on AI at a global level, however, does not fully respond to the doubts and critical issues regarding the training of algorithms and the creation of training datasets.

The contention between Italian DPA and OpenAI has highlighted how, in the absence of certain rules, it appears clear how the Supervisory Authorities must present themselves as interpreters of living law. However, although the work of the supervisory authorities themselves can highlight the problems and issues, in the resolution of the issue, it cannot have the systemic characteristics that a European or Statal regulation can have.

The theme of generative AI will probably be at the centre of debates in the coming years and will once again propose the dualism between "apocalyptic vs integrated", between those who fear the decline of the human species in the development of AI and those who, instead, welcome the advent of a new golden technological age. The next few months, with the outcome of the discussions between the Italian DPA and OpenAI and above all with the first applications of the AI Act, we will be able to better direct any reflection on the role of generative AI, understood as a risk or as an opportunity. At the same time, after the generative AI models are implemented there may be questions about the value of the law and the actual application of the AI Act to generative AI systems. Precisely in relation to this profile, therefore, the following paragraph focuses on some brief notes on the structure of the AI Act and on the rules that may affect generative models and provide protection against any risks arising from their use.

5 Generative AI Systems in the AI Act: "Generic" GPAI and "Systemic" GPAI[35]

As already highlighted,[36] in March 2024 the European Parliament in plenary session approved the text of the AI Act [37][37] with a very large majority, as the result of a complex itinerary that started in 2021 and as a first attempt at regulation expressly dedicated to AI, at a global level.

The purpose of the regulation is expressed in Art. 1 paragraph 1: "*to improve the functioning of the internal market and promote the uptake of human-centric and trustworthy artificial intelligence (AI), while ensuring a high level of protection of health, safety, fundamental rights enshrined in the Charter of Fundamental Rights, including democracy, the rule of law* and *environmental protection, against the harmful effects of artificial intelligence systems (AI systems) in the Union, and to support innovation*". Moreover, Art. 1 takes up the guidelines posed by Recital n. 1 which also expressly recalls the need for a unitary regulatory framework for AI, to harmonize legal regulation with the fundamental rules of the EU Charter and, at the same time, to protect the free movement of goods and services based on AI systems, while limiting the power of Member States to impose stricter rules [16].[38]

The approved text is also notable for choosing an approach based on the legal principle of prevention and on the classification of AI systems according to a risk scale, corresponding to a series of transparency and protection duties and, in the

---

[35] This paragraph was added in April 2024, after the approval of the AI Act by the EU Parliament; it replaces the part of the oral presentation dedicated to the proposal of EU regulation on AI.

[36] Please refer to paragraph 1 and notes 12 and 13.

[37] The adopted text can be consulted at web address https://www.europarl.europa.eu/doceo/document/TA-9-2024-0138_EN.pdf; among the first comments we note: C. Novelli, L'Artificial Intelligence Act (AIA) Europeo: alcune questioni di implementazione, in federalism.it, 2024, p. 95.

[38] In Recital n. 1 we can read "*The purpose of this Regulation is to improve the functioning of the internal market by laying down a uniform legal framework in particular for the development, the placing on the market, the putting into service and the use of artificial intelligence systems (AI systems) in the Union, in accordance with Union values, to promote the uptake of human centric and trustworthy artificial intelligence (AI) while ensuring a high level of protection of health, safety, fundamental rights as enshrined in the Charter of fundamental rights of the European Union (the 'Charter'), including democracy, the rule of law and environmental protection, against the harmful effects of AI systems in the Union, and to support innovation. This Regulation ensures the free movement, cross-border, of AI-based goods and services, thus preventing Member States from imposing restrictions on the development, marketing and use of AI systems, unless explicitly authorized by this Regulation*", while the successive Recital n. 2 provides that "*This Regulation should be applied in accordance with the values of the Union enshrined as in the Charter, facilitating the protection of natural persons, undertakings, democracy, the rule of law and environmental protection, while boosting innovation and employment and making the Union a leader in the uptake of trustworthy AI*". The regulation of AI systems is mainly aimed at market dynamics, however, as the European legislator himself highlights, it cannot neglect the level of fundamental rights and freedoms of the individuals: on the subject, see. F. Donati, Diritti fondamentali ed algoritmi nella proposta di Regolamento sull'Intelligenza Artificiale, in Il Diritto dell'Unione Europea, 2021, p. 453.

extreme case of unsustainable risk, corresponding to the absolute ban on AI system placings on the market and their diffusion [36].[39]

The complexity of the approved text and the multiplicity of relevant profiles of the new discipline would require a detailed analysis which, however, goes beyond the scope of this essay. In the following considerations, therefore, only some profiles will be analysed, relating to the regulation of generative AI systems which, moreover, constitutes an integration of the original text and was the subject of social and legal debates between EU Member States (mainly France and Germany), who were supporters of a lighter regulation to valorise the self-regulation of ICT Big Payers and more generally of providers, and the EU legislator who aimed, instead, at laying down binding duties. With regard to this latter position, the choice was for a regulation imposed by the EU institutions and for the centralization of a series of procedures and assessment regarding the ratings of the level of risks of the diffusion and use of AI systems, in the hands of the Commission.

Initial evidence of all amendments carried out and of attention toward technological evolution can already be found in the second paragraph of Article 1, indicating the general topics covered by the regulation, in letter (e) where we can read "*harmonised rules for the placing on the market of general-purpose AI models*", highlighting the relevance of computer models that can be used for the generation of content and the implementation of different tasks and functions.

Preliminary to any analysis on regulatory choices, however, we recall the definitions that the European legislator has chosen to use in the AI Act: Article 3 provides a rich range of definitions among which n. 1 stands out for its general definition of an AI system: "*a machine-based system designed to operate with varying levels of autonomy, that may exhibit adaptiveness after deployment and that, for explicit or implicit objectives, infers, from the input it receives, how to generate outputs such as predictions, content, recommendations, or decisions that can influence physical or virtual environments*". This is undoubtedly a very broad definition. However, this characteristic, which has been the subject of criticism by those who complained about a certain indeterminacy, could became a winning element if read in the light of the (probable) development of technology and the necessity to indicate the general category of every AI system model and algorithm that have new, innovative characteristics.

Scrolling through the 68 definitions of Article 3 of the AI Act, we find, at n. 63, the definition of General Purpose AI Model (GPAI) which is an "*AI model, including where such an AI model is trained with a large amount of data using self-supervision at scale, that displays significant generality and is capable of competently performing a wide range of distinct tasks regardless of the way the model is placed on the market*

[39] For an analysis of the proposed classification, see: G. Natale, Il nuovo Regolamento Europeo AI Act, in Diritto dell'Internet, 2024, p. 201; preliminary to the analysis of the main provisions of the Regulation, the author poses a series of questions that highlight some critical points of the AI Act and in particular underlines a doubt regarding the very general formulation of various provisions which lead to the question of whether we cannot essentially speak of a directive, albeit in the form of regulation, precisely due to the presence of various general clauses and the aim which appears to be the creation of a basis for a uniform law, rather than the realization of law harmonization.

and that can be integrated into a variety of down-stream systems or applications, except AI models that are used for research, de-development or prototyping activities before they are released on the market"; later, at n. 66, we find the definition of General Purpose AI System (GPAI-S) as "*an AI system which is based on a general-purpose AI model, that has the capability to serve a variety of purposes, both for direct use as well as for integration in other AI systems*". In more general terms, GPAI models are basic computer models (previously referred to as Foundation Models, FM) that require a large amount of data for training and, using machine learning settings, can then be used to carry out detailed tasks or as a basic element for the training of more complex specific AI systems aimed at particular tasks and functions. As an example of such models, we can consider the GPT itself which, connected to chatbot software, has demonstrated great ability to adapt and learn in the "communicative interaction in human language" environment, generating a series of complete and reliable texts both with the processing of content on request and with the implementation of specific activities, such as translations.

Attention to technological evolutions and consideration of the growing impact of the use of generative AI systems on the market has led the European legislator to dedicate special attention to GPAI models and GPAI systems, which are identified as generic models (GPAI) and systemic models (GPAI-S) and to which different risk categories are connected. In particular, systemic GPAI-S models are considered riskier and are therefore subjected to greater constraints than generic GPAI models.

As already mentioned, in the first draft of the AI Act, generative AI tools were not contemplated, probably due to their limited diffusion and the state-of-the-art. However, in the space of just two years, the provision of rules on generative AI and on LLM, those linguistic models trained on large quantities of data, has become of primary importance so much so as to determine the inclusion of two chapters in the AI Act through subsequent amendments and additions: these are chapter IV and chapter V, entirely dedicated to GPAI models and in particular to GPAI with systemic risk [17].[40]

Chapter IV is entirely dedicated to the transparency obligations that weigh on providers and deployers of certain AI systems, identified in Art. 50 as those "*systems capable of generating synthetic audio, image, video or text content*". In the case of contents created through these systems, a series of duties of disclosure are envisaged for providers and deployers who will have to place marks and labelling "in a machine-readable format" on the contents created through GPAI models to make them "detectable as artificially generated or manipulated". The provision of Art. 50 is mainly aimed at protecting the public who may be the recipient of content created

[40] For a definition of systemic risk, see Art. 3 no. 44: "*systemic risk means a risk that is specific to the high-impact capabilities of general-purpose AI models, having a significant impact on the Union market due to their reach, or due to actual or reasonably foreseeable negativity effects on public health, safety, public security, fundamental rights, or the society as a whole, that can be propagated at scale across the value chain*"; on this topic, see: D. Elgesem, The AI Act and the risk posed by generative models, in NAIS 2023: The 2023 symposium of the Norwegian AI Society, June 14–15, 2023, Bergen, Norway (https://ceur-ws.org/Vol-3431/paper3.pdf).

through generative AI, to prevent people from being deceived through the dissemination of deepfake content or artificially created texts. Alongside the protection of individuals, protection of the common interest also acquires importance, as is evident in the reference in paragraph 4.2, where the right to information on matters of public interest is taken into consideration and the declaration on the nature of the text generated by AI takes on the value of protection of the correct dynamics of public opinion shaping.[41]

The subsequent Chapter V is dedicated to GPAI models and, in Art. 51, indicates the criteria on the basis of which generic models can be considered systemic risk models, indicating the high impact capacity, as defined by Art. 3 no. 65,[42] as the decisive element. In particular, a GPAI system will be considered at systemic risk when it presents a high impact capacity on the basis of assessments carried out with "adequate technical tools and methodologies, including indicators and reference parameters" (as in letter a) or when there has been an evaluation carried out by the Commission acting as AI Office on its own initiative or "following a qualified report from the scientific group" (as in letter b). In this latter case, the Commission attributes the GPAI system "a capacity or impact equivalent to those referred to in letter (a), taking into account of the criteria set out in Annex XIII".

Furthermore, again in Article 51, paragraph 2 adds that high impact capability is assumed for GPAI systems "*when the cumulative number of calculations used for its training, measured in FLOPs, is greater than* 10^{25}".

From the terms of Art. 51 the central role of the Commission emerges, acting as AI Office (as provided for in Art. 3, n. 47) in the procedure for identifying GPAI systems at systemic risk defined by the subsequent Art. 52. In the presence of models and systems that have the requisites to determine or exclude systemic risk, indeed, it will be the Commission acting as AI Office that will evaluate positively (or not) the achievement of the requisites, the fulfilment of obligations/duties and the sufficiency of the results of any tests carried out and presented by the provider. In the case of GPAI systems, the precautionary approach, which characterizes the entire framework of the AI Act, is fully set out in the set of duties imposed on providers and deployers: firstly, carry out an evaluation of the GPAI model in compliance with protocols and standardized tools that reflect the state of technique advancement and also carry out forms of adversarial testing[43] to identify and limit systemic risk; then, carry out continuous monitoring activities to mitigate possible systemic risks and at the

[41] "*Deployers of an AI system that generates or manipulates text which is published with the purpose of informing the public on matters of public interest shall disclose that the text has been artificially generated or manipulated*".

[42] For the definition of high-impact capabilities, see Art. 3, n. 65: "*high-impact capabilities means capabilities that match or exceed the capabilities recorded in the most advanced general-purpose AI models*".

[43] The term "adversarial test" refers to a series of simulations aimed at testing the stability and computational capacity of the AI system through the use of inaccurate or harmful inputs, to verify the effect on machine learning methods; the so-called adversatives techniques, in fact, are generally known as methods aimed at distorting the functioning of AI systems precisely through the introduction of harmful inputs which determine the alteration of the output result.

same time promptly report the possible occurrence of serious incidents[44] to the AI Office and, in relation to them, adopt the most appropriate measures to restore the safety situation violated; lastly, permanently guarantee an adequate level of protection in terms of cybersecurity related to the model and the physical infrastructure of reference.

The fulfilment of the duties placed on the various generic and systemic GPAI models requires the definition of harmonized EU safety policies and compliance standards which, however, have yet to be defined and published. While awaiting this publication, Art. 52 provides that both categories of GPAI models can be based on "codes of practice"[45] that demonstrate the fulfilment of the legal imposed duties. In relation to the latter, however, it seems appropriate to remember that regarding the generic GPAI models, less stringent duties are envisaged, aimed mainly at guaranteeing transparency and accessibility to the technical documentation on functioning and methods of training.

The brief considerations carried out on chapters IV and V of the AI Act text approved by the European Parliament in March 2024 must be part of the broader field of reflection on the structure and implications of the AI Act which will come into force after its publication in the EU Official Journal, leaving Member States the usual time of 24 months to adapt national legislation before full applicability of the Regulation.

Several times, in the AI Act Recitals, the ideal of trustworthy AI is stated and confirmed, presuming that the ideal is also capable of gathering together respect for EU social values, the protection of citizens' rights and, at the same time, the interest of the digital market, so as to give the EU a leading role in this sphere. At the same time, it has not been possible to silence the doubts and perplexities of those who fear that, on the other hand, the proposed rules are too narrow and could constrain or hinder companies already present and those just entering the market.

The goal achieved by the EU, the first institution in the world to propose regulation on AI, undoubtedly marks a historic moment but at the same time presents a truly unprecedented challenge: being able to regulate a technological phenomenon whose evolution has proven to be frenetic and continuously accelerating over ever shorter time intervals. Furthermore, it must be able to keep that regulation valid also with regard to the most immediate technological developments and those coming soon in the next few years. The choice of the European legislator in using very general definitions, such as those of Art. 3, n. 1 (AI system) and nos. 63 and 65 (GPAI model and GPAI system), can be explained precisely by the need to project the AI Act beyond

[44] The last paragraph of Article 51 places the burden on the Commission to periodically update the thresholds indicated in the previous paragraphs, in consideration of "the evolution of technological developments, such as algorithmic improvements or greater efficiency of the hardware (...) so that these thresholds reflect the state of the art".

[45] The expression "codes of practice" is used to indicate a set of technical documents that report the standards of the reference in a technological sector; these documents must be developed by the providers and supervised by the AI Office, although formal approval of the same is not required. The presentation of the "code of practice" implies a presumption of conformity with the duties imposed by the AI Act, but does not represent a release of the model.

the technical contingency of the moment and to prevent obsolescence phenomena which the regulation could encounter following the most rapid evolutions.

Precisely in consideration of this unprecedented speed of time, technological evolution takes on a transcendent value of the contingent physical reality, while the legal values of the EU and the fundamental rights of individuals demonstrate their immanent value of that same physical reality, and indeed they constitute one of the so-called timeless "invariant characters of the Right": the choice to regulate AI therefore becomes the meeting and comparison ground of two different value modalities. The rule of positive law is asked once again and even more, in the context of the AI Act, to become a synthesis between technical rule and legal rule from the perspective of social evolution and contemporary transformations. And this could be the real challenge of the European Union's regulation on Artificial Intelligence: the historical record on the launch of AI regulation has been achieved, but the real and most efficient goal will be the resistance of this regulation to the slow passage of physical time and to the tumultuous unfolding of technological time.

References

1. Amato Mangiameli AC (2022) Intelligenza artificiale, big data e nuovi diritti. Riv Ital Inf Dir 93–101
2. Andreola E (2019) Minori e incapaci in Internet. ESI Editore, Napoli
3. Arnone C (2024) L'uso dei servizi di intelligenza artificiale da parte dei minori: solo una questione di "age verification"? Dir Internet 13–22
4. Battelli E (2022) Privacy e minori: l'inadeguatezza del consenso digitale. In: D'Auria M (ed) I problemi dell'informazione nel diritto civile, oggi. Studi in onore di Vincenzo Cuffaro. Ed. RomaTRE Press, Roma, pp 543–562
5. Bedessi S (2023) Intelligenza artificiale e fenomeni sociali. Ed. Maggioli, Sant'Arcangelo di Romagna
6. Bifulco L (2018) Intelligenza artificiale, Internet ed ordine spontaneo. In: Pizzetti F (ed) Intelligenza artificiale, protezione dei dati e regolazione. Ed. Giappichelli, Torino, pp 383–400
7. Bravo F (2017) Il consenso e le altre condizioni di liceità del trattamento dei dati personali. In: Finocchiaro G (ed) Il nuovo regolamento europeo sulla privacy e sulla protezione dei dati personali. Ed. Zanichelli, Bologna, pp 101–177
8. Cardarelli F (2021) Articolo 58. Poteri. In: D'Orazio R, Finocchiaro G, Pollicino O, Resta G (eds) Codice della Privacy e Data Protection, pp 741–752
9. Cascione CM (2021) Art. 8—Condizioni applicabili al consenso dei minori in relazione ai servizi della società dell'informazione. In: D'Orazio R, Finocchiaro G, Pollicino O, Resta G (eds) Codice della Privacy e Data Protection. Ed. Giuffrè, Milano, pp 229–239
10. Caselli A (2022) Art. 58—Poteri. In: Riccio GM, Scorza G, Belisario E (eds) GDPR e normativa privacy. WKI, Milano, pp 596–615
11. Citarella G, Venchiarutti A (2022) Diritti della personalità del minore e tutele nella rete Internet. In: D'Auria M (ed) I problemi dell'informazione nel diritto civile, oggi. Studi in onore di Vincenzo Cuffaro. Ed. RomaTRE Press, Roma, pp 527–542
12. D'Alfieri D (2022) Internet: quando la "rete" cattura i minori. Riv Ital Inf Dir 53–61
13. D'Aloja A (2019) Il diritto verso il "mondo nuovo". Le sfide dell'Intelligenza Artificiale. BioLaw J Riv Biodir 3–31
14. D'Avack L (2020) La rivoluzione tecnologica e la nuova era digitale: problemi etici. In: Ruffolo U (ed) Intelligenza artificiale. Il diritto, i diritti, l'etica. Ed. Giuffrè, Milano, pp 3–28

15. De Franceschi A (2021) Art. 4—Definizioni. In: D'Orazio R, Finocchiaro G, Pollicino O, Resta G (eds) Codice della Privacy e Data Protection. Ed. Giuffrè, Milano, pp 156–176
16. Donati F (2021) Diritti fondamentali ed algoritmi nella proposta di Regolamento sull'Intelligenza Artificiale. In: Il Diritto dell'Unione Europea, pp 453–466
17. Elgesem D (2023) The AI Act and the risk posed by generative models. In: NAIS 2023: the 2023 symposium of the Norwegian AI Society, Bergen, Norway, 14–15 June 2023. https://ceur-ws.org/Vol-3431/paper3.pdf
18. Faini F (2019) Big data, algoritmi e diritto. DPCE on line, pp 1869–1882
19. Ferrari V (2020) Note socio-giuridiche introduttive per una discussione su diritto, intelligenza artificiale e big data. Sociol Dir 9–32
20. Finocchiaro G (2017) Il nuovo Regolamento europeo sulla privacy e sulla protezione dei dati personali. Ed. Zanichelli, Bologna
21. Finocchiaro G (2022) La proposta di regolamento sull'intelligenza artificiale: il modello europeo bassato sulla gestione del rischio. Il Diritto dell'informazione e dell'informatica, pp 303–322
22. Finocchiaro G (2020) Riflessioni su Intelligenza artificiale e tutela dei dati personali. In: Ruffolo U (ed) Intelligenza artificiale. Il diritto, i diritti, l'etica, Milano, pp 237–250
23. Flordi L (2022) Etica dell'Intelligenza artificiale. Sviluppi, opportunità, sfide. Ed. Raffaello Cortina Editore, Milano
24. Galli G, Sartor F (2022) L'utilizzo dei Big Data e dell'IA per una migliore qualità della regolamentazione. In: Osservatorio sulle fonti, pp 34–360
25. Iacovelli D, Fontana M (2022) Nuove sfide della tecnologia e gestione dei rischi nella proposta di regolamento europeo sull'intelligenza artificiale: set di training, algoritmi e qualificazione dei dati. Profili critici. Il Dir Econ 107–138
26. Lagioia F, Sartor G, Simoncini A (2021) Art. 22—Processo decisione automatizzato relativo alle persone fisiche, compresa la profilazione. In: D'Orazio R, Finocchiaro G, Pollicino O, Resta G (eds) Codice della Privacy e Data Protection. Ed. Giuffrè, Milano, pp 378–390
27. Liguori L (2021) Informazione e accesso ai dati personali. In: D'Orazio R, Finocchiaro G, Pollicino P, Resta G (eds) Codice della Privacy e Data Protection. Ed. Giuffrè, Milano, pp 287–304
28. Maccabiani N (2022) Il percorso europeo verso la qualità dei dati e la sua standardizzazione nell'IA: una prospettiva legale. BioLaw J Riv Biodir 473–502
29. Malaschini A (2023) ChatGPT e simili: questioni giuridiche ed implicazioni sociali. Consulta online, 583–606
30. Manzotti R, Rossi S (2023) Io&Ia. Mente, cervello e GPT. Ed. Rubettino, Soveria Mannelli
31. Marchetti B, Parona L (2022) La regolazione dell'Intelligenza artificiale: Stati Uniti e Unione Europea alla ricerca di un possibile equilibrio. DPCE on line, pp 237–252
32. Mazzei G (2023) "Big Data" come beni comuni globali e principi in tema di "public utilities". In: federalismi.it, pp 162–184
33. Megale L (2023) Il Garante della privacy contro ChatGPT: quale ruolo per le autorità pubbliche nel bilanciare sostegno all'innovazione e tutela dei diritti? Giornale di diritto amministrativo, pp 403–413
34. Messinetti R (2021) Comunicare nell'infosfera. La vulnerabilità della persona digitale. Editoriale. In: federalismi.it, pp IV–XVII
35. Monreale A (2020) Rischi etico-legali dell'Intelligenza Artificiale. DPCE on line, pp 3391–3398
36. Natale G (2024) Il nuovo Regolamento Europeo AI Act. Dir Internet 201–210
37. Novelli C (2024) L'Artificial Intelligence Act (AIA) Europeo: alcune questioni di implementazione. In: federalismi.it, pp 95–113
38. Peluso MG (2023) Intelligenza artificiale e tutela dei dati. Ed. Giuffrè, Milano
39. Pesci G (2022) Prospettive e recenti sviluppi della tutela dei minori on line. Ciberspazio Dir 311–333
40. Piemonte J, Papakostantinou V (2023) Il caso Clearview AI: uno stress test per il Regolamento generale per la protezione dei dati e la proposta di Regolamento sull'intelligenza artificiale in relazione alle nuove sfide poste dal riconoscimento facciale. Cibespazio Dir 43–59

41. Pilla F, Cupini G, De Santo M, Di Maio M, Dolce R (2023) Uno, Nessuno, ChatGPT. Ed. Ledizioni, Milano
42. Poletti D (2021) Art. 6—Liceità del trattamento. In: D'Orazio R, Finocchiaro G, Pollicino O, Resta G (eds) Codice della Privacy e Data Protection. Ed. Giuffrè, Milano, pp 191–205
43. Prosperetti E (2023) ChatGPT ed altri tool di AI. Condizioni di utilizzo, questioni di proprietà intellettuale e privacy. Dir Internet 245–254
44. Prosperetti E (2024) La disciplina contrattuale dell'uso dell'IA generativa. Una lettura delle nuove condizioni d'uso dei servizi di OpenAI. Dir Internet 211–218
45. Raffiotta E, Baroni M (2022) Intelligenza artificiale, strumenti di identificazione e tutela della identità. BioLaw J 165–179
46. Resta F (2022) Art. 6—Liceità del trattamento. In: Riccio GM, Scorza G, Belisario E (eds) GDPR e normativa privacy. WKI, Milano, pp 78–94
47. Resta G (2022) Cosa c'è di "europeo" nella Proposta di Regolamento UE sull'intelligenza artificiale? Il Dir Inf Inf 323–342
48. Sapienza S (2023) ChatGPT, dati e identità personale: rischi, bilanciamenti, regolazione. Osserv costit 47–71
49. Silvetti V (ed) (2016) I provvedimenti in materia di data protection dei Garanti Europei. Ed. Giuffrè, Milano.
50. Spoto G (2021) Disciplina del consenso e tutela del minore. In: Sica S, D'Antonio G, Riccio GM (eds) La nuova disciplina europea della privacy. Milano, pp 111–130
51. Todaro P (2022) Art. 13—Informazioni da fornire qualora i dati personali siano raccolti presso l'interessato. In: Riccio GM, Scorza G, Belisario E (eds) GDPR e normativa Privacy, II edn. Milano, pp 175–205
52. Tresca M (2021) "Big data", "open data" e algoritmi. Riv trim dir pubbl 545–557

Predictability and Calculability in Tax Proceedings: The State of the Art

Maria Vittoria Serranò

Abstract Automatisms in statement audits represent one of the most widely used forms of artificial intelligence in the field of taxation. Issues related to balancing the right to protection of taxpayers' data and the fight against tax evasion also appear to be significant, which requires the use of increasingly advanced technological tools and will be the subject of critical reflection. With particular regard to the tax process, serial cases and the significant amount of appeals, provide the incipit for being able to introduce, refine and enhance the use of predictive justice. In the field of taxation, especially in the presence of serial cases and mass litigation (cadastral assessments, assessments by means of sector studies, income tax assessments, and so on), the precedent may lead to greater predictability for judgments in the making, without necessarily leading to a depersonalisation of judgments.

Keywords Intelligence · Artificial · Automatisms · Proceedings · Process

Good morning, everyone. I'd like to begin by extending my heartfelt gratitude to Prof. Domenico Marino for inviting me to participate in this significant two-day event, organised by the Department of Heritage, Architecture, and Urbanism (PAU) and the Centre for the Study of Economic and Territorial Policies, which features an impressive lineup of speakers. I will start by informing you that the limited time available to discuss the essential aspects of AI application in Tax Law will be divided with Professor Santa De Marco, who will address the procedural elements related to tax assessment (including the selection of at-risk taxpayers, audit methods, and monitoring of "anomalous" tax declarations, among others).

Regarding the topic assigned to me, and aligning with the conference's focus on technological evolution and digital implementation from a geopolitical perspective, I believe it is necessary to bring my testimony, not only as a university professor, but above all, as a tax judge witnessing a momentous turning point in procedural tax law.

M. V. Serranò (✉)
Messina University, Messina, Italy
e-mail: Maria.serrano@unime.it

D. Marino and M. Monaca (eds.), *Generative Artificial Intelligence and Fifth Industrial Revolution*, Lecture Notes in Networks and Systems 880,
https://doi.org/10.1007/978-3-031-73880-7_10

I aim to update you on the current state of tax processes and artificial intelligence, which are the main focus of these brief reflections.

Tax law constitutes one of the areas in which artificial intelligence can be most profitably applied, especially given its vast array of evolving primary norms, implementing provisions, administrative and judicial guidelines, which often shift interpretatively compared to previous ones. This dynamic nature of law and the structural crisis it is undergoing, particularly tax law, allows for a significant role to be attributed to legal calculability in the strictest sense. Originally a sociological concept by Max Weber, "legal calculability" was seen as a cornerstone of Western capitalism, offering predictability and standing in contrast to the peril of an "incalculable right." This idea of a predictable, "calculable" right is not new and it should be kept quite distinct from the notion that "resolving disputes means ending them, clarifying factual uncertainties and interpretive ambiguities, and choosing one solution among many. Decision-making in law isn't arbitrary; it arises from a measured comparison between the law's provisions and the actual events that have taken place … Montesquieu Baron, in "Esprit des Lois" (XI, chap. VI), famously said that judges are merely the mouthpiece of the law, unable to lessen its force or severity, thus emphasising the impartiality and fairness in legal judgment. Though often ridiculed by overzealous or unsophisticated novelists, this notion aligns with the idea that a judgment transforms the "words of the law" into "words of judgment," determining cases in an "inanimate," or objective and impersonal manner. The logical circle, as discussed previously, guarantees the calculability of judicial decisions which depend on the comparison between legal frameworks and specific facts."

Recently, the Constitutional Court's judgment no. 94/2023 on repeated recidivism and legal automation in the criminal justice system also addressed calculability. It noted that "the predictability of law, grounded in legislative discretion, responds to an essential need: it's calculability. However, a legal system based on (flexible) principles rather than (rigid) rules risks irreversibly undermining this calculability. The court also recognised that while some judgments appear to be driven by a quest for justice rooted in human solidarity and dignity, as expressed in the first paragraph of Article 3 of the Italian Constitution, this does not invalidate them. This reflects the inherent dilemma in the law itself: it serves as both a balance and a labyrinth, striving for impartiality and equilibrium, yet creating a complex web of logic. The struggle to define justice, whether it's religious, secular, or based in solidarity, remains an ongoing, unresolved challenge …"

The assertion that these "heretical" judgments are driven not by a quest for divine or enlightened justice, but rather by a pursuit of justice founded on human solidarity and dignity, is indeed compelling. This passage delves into the fundamental dichotomy inherent in the concept of law itself: its endeavour to mediate between conflicting interests while simultaneously weaving a complex web of rationale. The assertion highlights how these judgments are significant in their effort to align legal justice with the social dignity owed to every individual, a principle encapsulated in Article 3 of the Italian Constitution. This reveals the intrinsic paradox and intractable complexity at the heart of law itself. While laws strive to be neutral and establish balance, they also create intricate and sometimes perplexing logical structures that can confound as much as clarify. This ongoing struggle to define justice, whether it is based on religious, secular, or humanitarian principles, remains a persistent and unresolved challenge.

A legal system founded on principles inevitably requires interpretative spaces, sometimes even chaotic ones, because these principles are indeterminate and adaptable and therefore need to be interpreted. This necessitates a balance between the need for calculability (in order to ensure equality of all citizens in the eyes of the law and to allow each person to base their behaviour on the predictable consequences of their actions), and the requirement to honour constitutional promises that open the legal system to interpretation and evolution, rather than static stability.

These brief reflections emphasise the significance of Parliament's recent enactment of the 'Delegation to the Government for Tax Reform' by bringing attention to the issues of legal calculability and predictability. This pivotal legislation directs the executive to issue one or more legislative decrees within 24 months, including those "*to prevent and reduce tax evasion and avoidance, 1) by fully leveraging data from the tax registry's information system, enhancing risk analysis, and employing digital technologies and artificial intelligence solutions, in compliance with the European Union's personal data protection regulations (…) and to "augment the use of digital technologies, including that of deploying artificial intelligence systems, to ensure the availability and timely use of relevant information through complete interoperability between databases.*"

The National Recovery and Resilience Plan (PNRR), which specifically identifies digitisation of the Public Administration sector as a primary objective to revive the economy severely affected by the COVID-19 pandemic, undoubtedly provides decisive encouragement to utilise opportunities for technological innovation. However, the current legislation on investigatory procedures, assessment, and collection of taxes, which allows for wide scope application of automation, does not seem sufficiently attentive to taxpayer guarantees, especially with the introduction and enhancement of Artificial Intelligence. Therefore, attention to the protection of citizens' rights as taxpayers now assumes paramount importance. The new legislative framework promoted by the Delegated Law for tax reform must also respect the other fundamental rights provided for by the Constitution, as well as the principles set out by the European Union and the European Court of Human Rights, and take into account that under no circumstances can the use of Artificial Intelligence lead to the infringement of the taxpayer's right to be informed about the functioning and use of the algorithm.

In this regard, ruling no. 2270/2019 of the Council of State clearly establishes how "*the use of "robotic" procedures cannot, however, be a reason for bypassing the principles that conform our legal system and regulate the conduct of administrative activity. In fact, the technical norm that governs each algorithm still remains a general administrative regulation, constructed by man and not by machine, to be then (only) applied by the latter, even if this is done exclusively*" and, again, that "*the multidisciplinary characterisation of the algorithm (a construct that certainly requires not only legal skills, but also technical, computer, statistical, administrative ones) does not dispense from the necessity that the "technical formula," which in fact represents the algorithm, be accompanied by explanations that translate it into the "legal ruling" underlying it and make it readable and understandable, both for citizens and for the judge.*"

Artificial Intelligence (AI) is undeniably a critical social and political issue today, bearing substantial ethical implications. Its application spans diverse sciences, from Medicine to Law, leading to concepts like predictive medicine and justice. These AI applications are not meant to substitute the work of doctors or judges but to augment their capabilities. For instance, it is estimated that over 33% of tasks currently performed manually by doctors could be automated with the use of AI systems not only providing effective diagnostic support but also generating significant health care cost savings. In the USA alone, experts predict that by 2026, the adoption of AI technologies could result in healthcare cost savings exceeding 150 billion dollars.

Without delving into excessive technicalities, by Artificial Intelligence we mean a series of systems capable of executing tasks typically associated with natural human intelligence. This is achieved through algorithms designed to autonomously improve performance based on the input data received. This approach is also evident in the digitisation of Law, particularly Tax Law. The apprehension among practitioners about the risk of depersonalisation, especially in procedural matters, due to new technologies is understandable. However, we believe the risk of a machine replacing a judge to be minimal, and we will briefly outline why the risk of depersonalisation is unlikely.

An essential, undeniable aspect of AI implementation is the digitization of processes. Digitisation and AI are interconnected and can be considered two sides of the same coin: where one cannot contemplate effective AI use without first having completed the entire digitisation of the process through the computerisation of tasks and obligations that were previously exclusively managed in paper format.

The increasing volume of litigation in Tax Courts and the Supreme Court of Cassation (with an estimated 65% of pending appeals in Cassation in 2025 being directly related to tax matters), has dictated the need for a serious and radical reform of tax justice, which has been long desired by all the stakeholders involved in taxation and tax policy. This need aligns with the objectives of the National Recovery and Resilience Plan, which Parliament approved and the European Commission positively received. A key goal, among others, is the structural reform of tax justice, as implemented by Law No. 130/2022.

Furthermore, the realisation of a predictive system of tax justice also involves creating a database for predictive tax jurisprudence purposes so as to allows judges, lawyers, policymakers and other legal professionals to have access to data that will allow them to assess the likelihood of winning or losing a case, thereby reducing the chances of pursuing futile legal actions that have little chance of success. This concept is commonly known as legal calculability—the ability to measure and foresee ex ante litigation outcomes.

In this context, the Prodigit project (a strategic project for our country as part of the PNRR plans), is critical. It focuses on digitising the Tax Justice system and has received strong support from the Council of Presidency of Tax Justice. Collaborating entities include the University of Messina, the Ministry of Economy and Finance, and Sogei, who have actively worked on processing data related to tax meters, sector studies, and now on the synthetic indicators of tax reliability.

At this point, it remains unclear how the AI system will function and how it will align with judicial intentions. Central to its operation is the frequency is the frequency with which certain words or sequences of words appear in texts. Through sophisticated analysis of stored corpora, the system determines the predictability of case outcomes (which, among other things, encourages avoiding unnecessary lawsuits) and could restrict litigation to more reasonable parameters. Since 2019, the Electronic Tax Trial (PTT) platform has been mandatory, requiring document submission via SIGIT. Additionally, since 2022, tax rulings have been natively digital, totalling about 200,000 per year. This massive influx of tax data allows for the introduction of predictive justice through algorithms that aim at ensuring (as per Article 65 of the Judicial System) "strict observance and consistent interpretation of the law, Unity of the objective national law". This is especially relevant for so-called serial lawsuits where one of the two parties is always the same, either the tax agency Ade or taxpayer Ader.

The tools through which the AI system will operate are essentially ChatGPT-4, and more recently, Bard and Anthropic, which allow for the selection of precedents to achieve a predictive outcome.

As a tax judge, I feel the need to share a final thought, for example, on the determination and quantification of litigation expenses, which are settled based on ministerial tables. As a tax judge, I see that a significant portion of litigation, particularly in later stages, concerns litigation costs and determining the admissibility of appeals based on deadlines. In this regard, alert systems could check for notifications that miss deadlines. This could help address the backlog under discussion.

In summary, the concepts of predictability, predictiveness, and calculability are central to the comprehensive computerisation of the tax process.

Bibliography

1. Carleo A (2017) Calcolabilità giuridica. Il Mulino
2. Irti N (2016) Un diritto incalcolabile. Giappichelli, Torino
3. Sulla rotta di un "diritto del caso singolo": la Consulta tra calcolabilità giuridica, discrezionalità, esigenze di giustizia (nota a Corte cost., sentenza n. 94/2023) di Nikolaj Vinai in OSSERVATORIO COSTITUZIONALE, Fasc. 6/2023
4. I dati riportati sono stati pubblicati in Il Sole 24 Ore del 3 maggio 2022
5. On the topic of the relationship between tax litigation and artificial intelligence, see Mastroiacovo V (2023) Predictability, predictivity and humanity in tax litigation judgment. Riv Telemat Dir Tribut; Contrino A (2023) Evolutionary pressures (at the supranational level) and involutionary pressures (at the internal level) on the balance between the right to protection of taxpayer data and the needs to combat tax evasion. Riv Telemat Dir Tribut; Marello E (2022) Update on... popper, prodigit and predictive justice. Riv Telemat Dir Tribut

The Applications of Artificial Intelligence in Automated Checks of Tax Declarations. Critical Reconstructive Insights

Santa De Marco

Abstract Automatisms in statement audits represent one of the most widely used forms of artificial intelligence in the field of taxation. In fact, the legislature, with the Delegated Law N. 111/2023, enhances the tools to support the activity carried out by the Tax Administration, where, artificial intelligence, machine learning and text mining, strengthen the quality of controls, aimed at identifying even the taxpayers at greater tax risk and at the same time simplify the assessment process. It should be pointed out that, frequently, the automatism of Artificial Intelligence generates a mismatch between actual and potential income. In order to protect the rights of the taxpayer, it is necessary to achieve a new perspective on the use of algorithms in procedural and assessment activities.

Keywords Intelligence · Artificial · Proceedings

1 Introduction

Tax law, by virtue of its increasingly mechanistic evolution and the enormous amount of data collected from different sources, is perfectly suited to the use of artificial intelligence and exerts its effects in three major directions: taxes, relations between the Tax Office and taxpayers and the process. With this contribution, of the different uses of artificial intelligence, we would like to focus our attention on the automatisms in statement audits which, also, represents one of the most used forms of artificial intelligence in tax matters.

The application of artificial intelligence in the realm of automated and formal controls is intriguing, to say the least, as it introduces it in the procedural phase,

S. De Marco (✉)
Messina University, Messina, Italy
e-mail: santa.demarco@unime.it

D. Marino and M. Monaca (eds.), *Generative Artificial Intelligence and Fifth Industrial Revolution*, Lecture Notes in Networks and Systems 880,
https://doi.org/10.1007/978-3-031-73880-7_11

where legislation must strike a balance between fiscal interest and equitable taxation, navigating the intricate relationship among governmental power, individual freedoms, and taxpayer rights.

It prompts a critical question as to the extent to which the integration of AI in the initial vetting stage, "on an already barren terrain in terms of democracy, could tilt the balance further from taxpayer rights and more towards authoritative power".

Delegated Law No. 111/2023, in various sections, anticipates the use of artificial intelligence (AI), assigning it a pivotal role in combating tax evasion and avoidance, as well as fostering a culture of voluntary compliance among taxpayers. While acknowledging the substantial benefits AI offers to both taxpayers and the Tax Administration, it's crucial to address the significant challenges posed by AI-driven audit controls. These challenges are evident in income tax audits as per Article 36-bis of Presidential Decree 600/73, and VAT audits under Article 54-bis of Presidential Decree 633/72. The concerns revolve around ensuring taxpayer protection, adhering to constitutional principles related to actual and potential income, and in relation to the knowability of the AI algorithms' programming.

2 The Operation of Artificial Intelligence in Automated Controls

The legislature with Delegated Law No. 111/2023 significantly enhances the resources available to the Tax Administration to carry out their duties. It introduces advanced technologies like artificial intelligence, machine learning, and text mining to improve the quality of tax controls aimed at identifying taxpayers with higher tax risks while also streamlining investigative processes and assessment procedures.

While various algorithm types (including non-deterministic, machine learning, and probabilistic models), can be applied, we find deterministic algorithms to be the most reliable for accurate tax reconstruction since the outcomes they produce are directly and transparently determined by the input data.

When we examine the automated audits of tax returns (specifically under Articles 54-bis of Presidential Decree 633/72 for VAT and 36-bis of Presidential Decree 600/73 for income taxes), it's important to first acknowledge that the control process applied to taxpayer submissions undergoes various stages: automated controls, formal controls, and substantive controls. Specifically, through automated checks or assessments, the tax administration employs specialised software to analyse the data in the tax registry alongside the information provided in the taxpayer's declaration (as per Articles 54-bis and 36-bis). This process enables the administration to recalculate the tax owed, correct computational or factual errors in tax credit allocations or deduction claims, and verify the alignment between the taxes due and the amounts paid by the taxpayer.

Typically, the result of this audit is communicated to the taxpayer via a so-called 'warning notice.' This notice allows the taxpayer to present their arguments or correct the identified discrepancies, possibly qualifying for a reduced penalty.

The introduction to the regulations governing the audit of income tax returns under Article 36-bis of Presidential Decree 600/73, and VAT under Article 54-bis of D.P.R. 633/72, both explicitly mention the use of "automated procedures." These automated checks are invaluable in assisting the Offices with the scrutiny of a large volume of tax declarations.

Significantly, these automated assessments replace traditional evaluations, since, the Tax Administration, via the use of advanced computer systems, recalculates the tax obligations (both the fact and amount of the debt) based on the data provided by the taxpayer and that available in the Tax Registry.

In such cases, this process quickly corrects any errors, leading to a re-evaluation of the credits or debts declared by the taxpayer. However, it's important to note that these recalculations may not always accurately reflect the actual credits or debts accrued by the taxpayer.

This process often bypasses the issuance of a standard assessment notice, to the extent that the audit directly results in a record, which might not always be accompanied by a notification of irregularity (commonly known as a 'remedial notice'), aimed at prompting the taxpayer to amend their submission.

This method of inspection, which essentially replaces traditional assessment, encounters its own set of challenges. These include internal limitations (such as correcting material mistakes made by taxpayers in the carryover of tax excesses), and external constraints, namely the boundaries within which the activities peremptorily stipulated in the aforementioned articles can be carried out.

On a purely practical level, we are witnessing an expansion in the scope of tax liquidations, attributed to various factors, including a liberal interpretation of clause 2(a) in Articles 36-bis and 54-bis. These articles state that 'material and calculation errors made by taxpayers in the determination of turnover and tax may be corrected.' However, the enumeration of taxpayer errors is neither exhaustive nor precise. Despite being a paper-based audit, a notable deficit in transparency exists, primarily because the algorithm from which the computerised audit arises remains undisclosed.

This situation raises an important question: how effectively can we reconcile the demands of digitalisation with safeguarding taxpayer rights? Specifically, issues like ensuring the pre-audit adversarial process, guaranteeing the accessibility and knowability of administrative acts, and maintaining clarity, motivation, transparency, cooperation, and good faith in these processes are of paramount concern.

3 Automated Audits and Protection of the Taxpayer

The audits under review are essentially the manifestation of regulatory data interpreted by a computer system programmed with specific algorithms. However, the limited scope of regulations inevitably impacts the programming of these algorithms.

Despite this, case law, through numerous rulings, has highlighted how the employment of various technological tools must adhere to certain principles, including knowability. A notable example is the Council of State's ruling No. 2270/2019, which states that 'the technical formula, essentially the algorithm, must be accompanied by explanations that convert it into understandable legal language.' This is crucial to ensure that it is comprehensible to both citizens and judges.

Furthermore, the Ministerial Decree of June 28, 2022, introduced a provision for human intervention to safeguard taxpayer interests. However, in practice, the legal framework does not seem to offer adequate protection against the challenges posed by digital audits and automated assessments. A frequent issue observed is the disparity created by AI's automatic processes between the actual and potential income of taxpayers.

In essence, the specific criteria underlying the calculation of results in tax assessments are not known, nor knowable. This is exemplified by the Synthetic Reliability Indices, used to gauge the tax reliability of professionals and businesses. In these cases, the basis for determining results is neither known, nor knowable, often leading to ambiguous outcomes during taxpayer simulations.

It's reasonable to infer that mass profiling of taxpayers, conducted through algorithms with unverifiable mechanisms, can directly influence the issuance of assessment notices.

The same considerations can also be made for tax settlements. However, unlike the Indirect Synthetic Acceptability Indices (ISAs), automated settlements under Articles 36-bis and 54-bis immediately trigger the recovery of dues. This process is initiated with the issuance of a payment without being able to make use of the end procedural cross-examination.

The numerous instances that underline the lack of transparency in the programming of IT systems for automated audits include the notable judgment by the Court of Cassation on April 7, 2022, case no. 11332. In this specific case, a taxpayer contested a payment notice under Article 36-bis. The dispute was regarding a higher tax (Ires for the year 2010), attributed to income from holdings in companies situated in tax havens with favourable tax regimes (Controlled Foreign Companies, CFC). The income in question was taxed based on transparency and subjected to separate taxation. The core issue was the application of a 27.5% tax rate, as opposed to the 27% rate claimed by the company. The company had applied the 27% rate—in line with the provisions of Article 167 TUIR at that time—as the base rate for income from its foreign subsidiary, citing the absence of income. However, the Supreme Court sided with the tax authorities, stating that the rate mentioned in the then-effective legislation was merely the minimum of the average rate. Therefore, in cases where the average rate was not applied, the standard rate of 27.5% was applicable.

Setting aside the specific interpretation of Article 167 TUIR by the Supreme Court, which necessitated a mere correction of the tax rate, it is important to highlight the frequent deficiencies in the programming of the control system. These deficiencies often result in what can be described as a "sort of digital assessment."

Based on a narrow interpretation of paragraph 3 in articles 36-bis and 54-bis, in the event the liquidation outcome differs from the tax return, the Tax Administration typically proceeds with registration on the tax roll. This approach, endorsed by both case law and tax administration, suggests that communication is not required in cases of non-payment compared to what the taxpayer has already settled. However, it's crucial to note that not all instances involve omitted payments. The variety of cases is broad, ranging from technical issues where payments made using the F.24 form don't correspond with the tax return, to complications related to tax offsets.

It must be emphasised that, in such scenarios, the taxpayer's right to defence during the intra-procedural phase is compromised. This necessitates a clear understanding that while algorithms can assist in decision-making, they cannot and should not replace human judgment, especially considering the critical issues they pose in terms of liability and privacy protection of taxpayers.

Therefore, ensuring that all algorithm-based activities adhere to the overarching principles of administrative conduct as outlined in the Administrative Procedure Act and the Statute of Taxpayers' Rights is deemed vital.

Case law has consistently emphasised the importance of the algorithm's knowability in "computerised administrative acts." The algorithmic process must be accompanied by explanations that translate it into the judicial law underlying the tax assessment act. This is crucial to ensure that taxpayers are informed about the automated decision-making processes affecting them and are able to understand or contest both the fact and the amount of the debt.

In the context of our discussion, there's also a concern about constitutional compatibility. In the realm of taxation, enforcement is governed by the rule of law principle, as per Article 23 of the Constitution. This implies that the legal basis for taxation should stem from a primary rule, to prevent the determination of the pursued interests from being left to the discretion of the Inland Revenue Services, rather than being defined by the legislature.

De iure condendo, to address these challenges, it's essential to involve the taxpayer in the procedure. Currently, Legislative Decree No. 219/2023, implementing the delegated law on the statute of taxpayer's rights, introduces both systematic and operational complexities.

Notably, Article 6-bis of the statute of taxpayer's rights, introduced by Legislative Decree no. 219/2023, mandates that all independently appealable acts must be preceded by an informed and effective cross-examination. The absence of such a general cross-examination before issuing an assessment notice could invalidate the procedure. However, the decree includes exceptions, notably it will not be applicable for automated and formal inspections.

This raises significant concerns "about protecting taxpayer rights. Excluding the adversarial principle based on the technology used, rather than on the quality of evidence of evasion, reflects a troubling trend."

To safeguard taxpayer rights, it's imperative to reconsider the use of algorithms in procedural and assessment activities, emphasising taxpayer involvement. Recently, the adversarial principle has marked a significant shift in the tax legal relationship between the Tax Office and taxpayers, fostering greater fairness, transparency, certainty, and cooperation. Additionally, in the context of information exchange between financial authorities at the European and international levels, tax inspections should adhere to the principles of effectiveness, efficiency, and economy of administrative action, ensuring the activation of prior cross-examination to respect taxpayer rights.

4 Brief Conclusions and Future Prospects

Thus, in conclusion, acknowledging the challenges that automated controls pose to the protection of taxpayer rights, it is hoped that the legislator, guided by constitutional principles and key doctrinal and jurisprudential guidelines, will reconsider the legislation concerning the use of artificial intelligence in automated tax return audits.

This reconsideration should not overlook the potential that AI, including advanced models like ChatGPT, blockchain applications, and shared databases, offers. These technologies have the promise to quickly reduce or even eliminate discrepancies that that could result in a tax basis that differs from the real one. However, it is crucial to remember that AI should be seen as an auxiliary tool, supporting the primary and irreplaceable role of human judgment that is solely and exclusively the responsibility of human beings. In line with the European principle of non-exclusivity in algorithmic decision-making, Tax Officers must take responsibility for verifying the reliability of algorithm-derived results. The criteria used in these algorithmic processes must be must be known and knowable.

Furthermore, as mandated by Article 53 of the Constitution, tax authorities must consider the actual contributory capacity, aiming to reconcile the highest computed income with the genuine contributory capacity of those under audit and assessment, thereby respecting the fundamental connection between taxation and contributory capacity.

In evaluating the balance between the purpose of assessment tools and taxpayer rights, actions should be directed towards creating a more equitable and efficient Inland Revenue System. This necessitates "striking a fair balance between public (including fiscal) interests and those of private individuals, adhering to principles of reasonableness and proportionality."

It would be fitting to envisage an endo-procedural cross-examination even in automated audits of tax returns, which increasingly resemble digital assessments. Such a process would allow tax authorities to refine data from estimated to real, potentially resolving issues that often escalate to litigation during this procedure.

It is imperative to highlight that tax interests must align with the principles of contributory capacity, effectiveness, proportionality, and reasonableness. The incorporation of an obligatory endo-procedural cross-examination represents a significant advancement in protecting the rights of taxpayers subject to audits.

Bibliography

1. Guidara A (2023) Accertamento dei tributi e intelligenza artificiale: prime riflessioni per una visione di sistema. Dir Prat Trib 2:385
2. Giovannini A (2023) Note controvento su interesse fiscale e "giustizia nell'imposizione" come diritto fondamentale (muovendo dall'intelligenza artificiale). Riv Dirit Tribut 3:252
3. Russo P (1995) Il problema dei termini per la liquidazione delle imposte dovute in base alla liquidazione ai sensi dell'art. 36-bis, Rassegna Tributaria, 1014 pp
4. Zagà S (2015) Le discipline del contraddittorio nei procedimenti di <<controllo cartolare>> delle dichiarazioni. Dir Prat Trib 6:845
5. Del Federico L (2023) Il sistema tributario e l'economia digitale. In: Del Federico L, Paparella F (eds) Diritto Tributario Digitale. Pacini Giuridica, Pisa, p 43
6. Francioso C (2023) Intelligenza artificiale nell'istruttoria e nuove esigenze di tutele. Rassegna Trib 1:47
7. Farri F (2020) Digitalizzazione dell'Amministrazione finanziaria e diritti dei contribuenti. Riv Dirit Tribut 6:115
8. Ragucci G (2023) I diritti del contribuente. In: Contraddittorio generalizzato prima di emettere l'atto impositivo, I focus del sole 24 ore, 2 Nov 2023
9. Uricchio AF, Parente SA (2020) Data driven e digital taxaction: prime sperimentazioni e nuovi modelli di prelievo. Dir Prat Trib 2:606
10. Dorigo S (2020) Il ragionamento giuridico nell'era dell'intelligenza artificiale. Pacini Editore, Pisa, p 319
11. Serranò MV (2014) Il rispetto del principio di proporzionalità e le garanzie del contribuente. Riv Dirit Tribut 4:881

Simulation of the Impact of Digital Innovation on Export Volumes and CO_2 Emissions in Italy

Giuseppe Quattrone, Fausto Cavallaro, and Domenico Marino

Abstract The aim of this research is to elaborate a model for evaluative and predictive analyses of CO_2 emissions into the atmosphere generated by the economic and production system and at the same time to demonstrate how attention to the reduction of environmental impact can also be profitable from an economic point of view for companies that decide to implement solutions and technologies capable of activating virtuous processes of decarbonization of the global economy in order to achieve climate neutrality objectives. To this end, an analysis was conducted on the sectors that are the cause of the production of greenhouse pollutant emissions with particular reference to anthropogenic emissions, focusing on the transport sector, which accounts for a significant share in the production of pollutant emissions. Therefore, a simulation model was defined for the flow of goods transport produced in Italy and destined for foreign markets for which the relative CO_2 emissions were examined. The orientation held by this research falls on the so-called "Sustainable Logistics" aimed at identifying optimal solutions not only in economic-business terms, but also in terms of containing environmental impact, following the hypothesis that: a reduction in environmental impact can also be profitable from an economic point of view for companies.

Keywords Sustainability · Logistics · Internationalization · Transport

1 Rational of the Study

The aim of this work is to develop a model for evaluative and predictive analyses of CO_2 emissions into the atmosphere by the production system and, at the same time, to demonstrate how the focus on reducing environmental impact can also be

G. Quattrone (✉) · F. Cavallaro
University of Molise, Campobasso, Italy
e-mail: gquattrone1@gmail.com

D. Marino
Mediterranea University of Reggio Calabria, Reggio Calabria, Italy

D. Marino and M. Monaca (eds.), *Generative Artificial Intelligence and Fifth Industrial Revolution*, Lecture Notes in Networks and Systems 880,
https://doi.org/10.1007/978-3-031-73880-7_12

profitable from an economic point of view for companies that decide to implement solutions and technologies with positive effects on the decarbonization of the global economy and the achievement of climate neutrality.

To this end, a preliminary analysis was conducted on the sectors that are the cause of greenhouse pollutant emissions, with particular reference to anthropogenic emissions.

In addition, data from the National Inventory of Atmospheric Emissions of the ISPRA (Istituto Superiore per la Protezione Ambientale) relating to the time series of greenhouse gas emissions from 1990 to 2021 were collected and catalogued, on which an exploratory analysis was conducted in order to reconstruct the main descriptive results cited on the reports and identify further elements for further research.

From this first analysis, it was decided to focus attention on the transport sector, which represents a significant share in the production of polluting emissions and on which to set the construction of modelling for the evaluative and predictive analysis of emissions. In this research, transport will therefore constitute the element of service supply with respect to a market demand given by the volumes of goods produced in Italy and destined for foreign markets. It should be noted that, with respect to the study of transport-related models, the orientation held by this research falls on the so-called "Sustainable Logistics" aimed at identifying optimal solutions not only in economic-business terms but also in terms of limiting environmental impact, following the hypothesis that: a reduction in environmental impact can also be profitable from an economic point of view for companies [1–3].

A number of authors [4] had already confirmed this insight, testifying to the interest of the topics in question and to the fact that it is an extremely innovative field of research for which technology transfer to companies has still achieved little. Numerous scientific articles relating to the areas of interest described above [5–9] representative of the vast literature on so-called 'Green Logistics' have therefore been collected.

As far as modelling is concerned, the model to be adopted for evaluative analysis is the transport model typically used for modelling problems in various management areas such as resource scheduling, production planning, and the design of distribution networks, in both the goods-producing and service sectors [10–12].

In addition to scientific articles, the bibliographic work also focused on national and international technical journals in the sector and on the examination of numerous websites of environmental and pollution observatories [13–16], logistics and inter-modal transport companies, international trade associations, as well as research centers engaged in the construction of environmental impact indicators for use in the transport sector [17–19].

1.1 Member States' Agreement on Climate Change

In 2015, at the Paris Climate Conference, a new post-2020 climate agreement was adopted which, for the first time, binds all countries to reduce their greenhouse gas emissions. Thus, the distinction made until then between industrialized and developing countries was largely abolished.

The Paris Agreement is a legally binding instrument under the Climate Convention. It entered into force on 5 October 2016 after reaching a quorum of 55 countries, which generate 55% of global emissions.

The Paris Agreement contains elements for a gradual reduction of global greenhouse gas emissions and establishes for the first time common principles for all countries pursuing the goal of limiting average global warming to well below 2 °C compared to pre-industrial times, aiming for a maximum increase of 1.5°. In addition, it aims to direct financial flows towards low greenhouse gas emission development and to improve the ability to adapt to climate change.

The common principles of the Paris Agreement stipulate that:

- Every five years, all countries must submit and illustrate a nationally determined reduction target (Nationally Determined Contribution, NDC). The achievement of the targets is only politically binding. In contrast, the implementation of the national measures, the reporting on the targets and the related international verification are binding.
- The reduction targets of all countries must be clear, understandable and quantifiable. The next reduction target must depend on the previous one and be as ambitious as possible.
- For the attainment of targets under the Agreement, emissions achieved abroad are allowed as long as they are environmentally sound, contribute to sustainable development and do not cause double counting. Article 6 of the Paris Agreement allows two types of emission reductions achieved abroad (Internationally Transferred Mitigation Outcomes, ITMOS): those resulting from a mechanism regulated by the Paris Agreement (Art. 6.4) and those resulting from bilateral and multilateral agreements (Art. 6.2).
- The Agreement largely ends the distinction in the narrow sense between industrialized and developing countries. The poorest countries are given some leeway in implementation. Industrialized countries are urged to take the lead by setting absolute reduction targets for the economy as a whole. In return, developing countries are urged to progressively set targets on the economy as a whole. The distinction between countries is dynamic, as reduction targets are set at the national level and must represent the highest possible ambition of a country. Each country's reduction target is thus measured by its climate responsibility and capacity.
- As far as adaptation to climate change is concerned, all countries must develop, present and regularly update adaptation strategies and measures. Each country may define the timing and form of the presentation at the international level itself. Countries are also obliged to report regularly on adaptation measures. The agreement strengthens existing loss and damage (Loss and Damage) prevention and mitigation mechanisms, explicitly excluding liability and compensation.
- The Paris Agreement does not set any new obligations for climate financing. Industrialized countries are obliged, as hitherto, to support developing countries in taking their own measures to reduce emissions and adapt to climate change. For the

first time, non-industrialized countries have also been invited. The mobilization of public and private funds is now everyone's responsibility. However, industrialized countries must continue to play a pioneering role. The common target of the industrialized countries to mobilize USD 100 billion of private and public funds annually from 2020 onwards has been confirmed until 2025, while a new, similar target has been outlined for the period thereafter. Industrialized countries are thus still obliged to report every two years on the funds mobilized and, as far as possible, to provide indications on the quality and quantity of funds planned for the following years. Similarly, developing countries are required to report every two years on the funds needed and received as well as on the climate-friendly investments they have mobilized and on international climate financing.

1.2 Analysis of ISPRA Data on CO_2 $Emissions_{eq.}$ in Italy

CO_2 equivalents (CO_2 eq) are a unit of measurement needed to uniformly express the climate impact of different greenhouse gases. In addition to the main man-made greenhouse gas, carbon dioxide (CO_2), there are other greenhouse gases, including methane and nitrous oxide. The different gases do not contribute equally to the greenhouse effect and remain in the atmosphere for different periods of time. Methane for example has 28 times more climate impact than CO_2, but stays in the atmosphere for less time. The climate effect of nitrous oxide exceeds that of CO_2 by almost 300 times. Greenhouse gas emissions can therefore be converted to CO_2 equivalents and then grouped together.

An initial analysis was conducted on a national territorial scale using data from ISPRA's National Atmospheric Emissions Inventory relating to the time series of greenhouse gas emissions from 1990 to 2021 (the last year available).

In this first analysis, we limited ourselves to the study of the weight of CO_2 emissions in the different economic sectors classified according to the IPCC (Intergovernmental Panel on Climate Change)—UN Intergovernmental Panel on Climate Change—and their observed trends over the period 1990–2021.

If we analyze the data on CO_2 $emissions_{eq.}$ in Italy between 1990 and 2021 (Fig. 1), the observed trend shows an increasing trend of these atmospheric emissions until 2005, after which the trend tends to decrease steadily until it goes below the 90 levels starting in 2009. This behavior is observed both in the case of emissions calculated with LULUCF (Land Use, Land Use Change and Forestry) and without probably due to emission reduction measures adopted by the sectors, particularly in the energy and manufacturing industries.

If one breaks down the atmospheric emission inventory data according to the five main economic sectors: Energy, Industrial processes, Agriculture, Land use and Waste, the energy production sector, which includes the energy industries (Energy industries), the manufacturing industries and construction (Manufacturing industries and construction) and the transport sector (Transport), is the main contributor to CO_2 $emissions_{eq.}$ and accounts for an average of 79.7% of national emissions (Table 1).

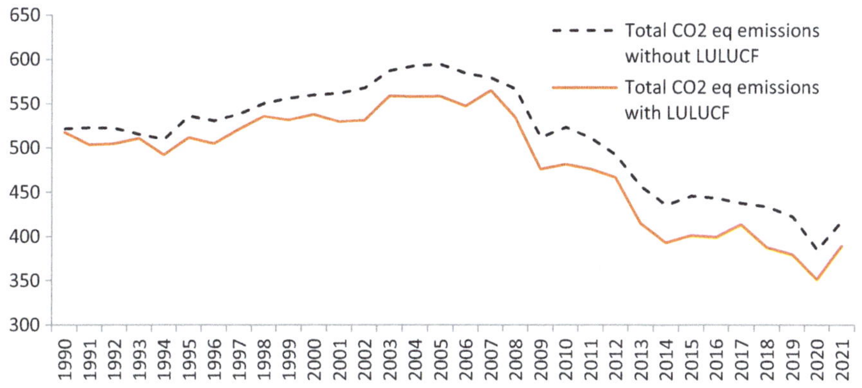

Fig. 1 CO_2 eq emission trend—year 1990–2021

Table 1 GHG CO_2 eq emissions by sector in Italy—value expressed in kton

Greenhouse gas source and sink categories	2021	%
Energy	332,832.13	79.7
Industrial processes and product use	31,852.11	7.6
Agriculture	32,717.22	7.8
Waste	20,189.97	4.8
Total (without LULUCF)	417,591.43	100

Source Our elaborations on ISPRA data 2021

Since 2005, GHG emissions from the energy sector (Fig. 2) have been decreasing due to policies adopted at European and national level to implement energy production from renewable sources. Since the same year, a further shift from oil products to natural gas in energy production has been observed as a consequence of the start of the EU greenhouse gas emission trading system (EU ETS) on 1 January 2005. Since 2009, a further decline in emissions is observed due to the economic recessions of 2008/2009 and 2012/2013. Ultimately, energy consumption from primary sources over the period 1990–2021 shows a clear shift from oil products and solid fuels to natural gas and renewables, while the share of electricity consumption is variable and market-driven.

If we look further at the level of emissions generated by the 'Energy' sector, to which the 'Transport' sub-sector belongs, the latter's contribution to CO_2 eq emissions is 24.7% of the total, and of this, almost 93% is road transport, which alone generates 95.7 Mtonnes of CO_2 eq into the atmosphere, and has also seen an increase in its contribution to emissions over the past 10 years.

A further deepening of the data on emissions from the transport sector allows us to break down the weight according to the type of vehicle used. In this case, passenger cars account for 64.5% of emissions followed by heavy goods vehicles (20.7%) and light trucks (11.9%) (Table 2).

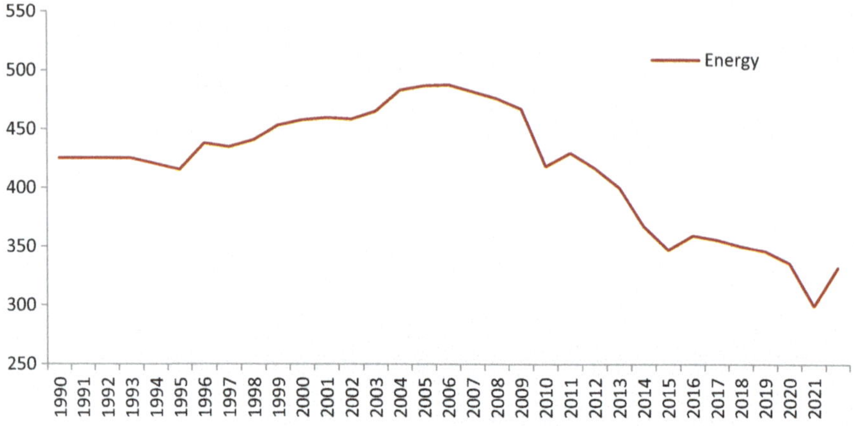

Fig. 2 CO_2 eq emission trend (energy sector)—year 1990–2021

Table 2 GHG CO_2 eq emissions by road transport mode in Italy—value expressed in kton

Transport mode	2021	%
Cars	64.34	64.5
Light duty trucks	11.90	11.9
Heavy duty trucks and buses	20.70	20.7
Motorbikes	2.83	2.8
Other (please specify)	0.00	0.0
Total road transportation emission	99.77	100

Source Our elaborations on ISPRA data 2021

It is evident from these results that the high demand for road transport for people and goods makes the problem of road transport pollution particularly relevant, especially in urban areas, due to the impact on pollutants such as carbon monoxide, volatile organic compounds (including benzene), nitrogen oxides and particulate matter, as well as the significant contribution to greenhouse gas emissions, particularly carbon dioxide.

1.3 CO_2 Emissions on a Regional Territorial Scale

With regard to CO_2 emissions on a regional territorial scale, an analysis was carried out on data from the Provincial Air Emissions Inventory. Air emissions database for the year 2019, classified by CORINAIR activity level (SNAP), disaggregated from the national inventory using top-down methodology.

The disaggregation of the national inventory at provincial level is carried out for all emission sources (natural and anthropogenic), subdivided according to the nomenclature used at European level EMEP/EEA, which classifies activities according to SNAP by grouping them into 11 main macro sectors:

1. Combustion—energy and the processing industry,
2. Combustion—non-industrial,
3. Combustion—industry,
4. Production processes,
5. Extraction, distribution of fossil/geothermal fuels,
6. Use of solvents,
7. Road transport,
8. Other mobile sources,
9. Waste treatment and disposal,
10. Agriculture,
11. Other emission and absorption sources.

For this purpose, it was deemed necessary to reconstruct the 11 SNAP'97 macro-sectors (Selected Nomenclature for Air Pollution) as the SNAP variable was relative to individual sectors. This was made possible by cross-referencing the data from the dataset in question with the SNAP'97 classification given in the 'Guidelines for local air emission inventories' published by the National Agency for Environmental Protection ANPA in 2021.

On a regional level, the largest CO_2 emissions are generated by the region of Lombardy (20.4%), followed by Emilia Romagna (10.3%), Apulia (10%) and Veneto (9.7%) (Table 3).

CO_2 emissions from road freight transport account for a significant share of total GHG emissions in Italy. In 2019, the transport sector contributed 25.2% of total national GHG emissions, with 92.6% of emissions coming from road transport. In particular, road freight transport is responsible for a significant share of CO_2 emissions, with an increase of 3.9% from 1990 to 2019.

Looking at the emissions figures for the road transport sector alone on a regional territorial scale, Lombardy retains the top spot with emissions accounting for 16.9% of the national total, followed by Lazio (9.6%), Emilia Romagna (9.4%) and Veneto (9.3%) (Table 4).

It should be noted that the COPERT model (Computer Programme to calculate emissions from road transport) whose technical development is supported by the European Environment Agency as part of the activities of the ETC/ACC (European Topic Centre on Air and Climate Change) used by ISPRA to estimate pollution from road traffic is a national methodology developed in accordance with guidelines and good practice. This model is fundamental for assessing emissions from road vehicles and contributes to the drafting of the national inventory of atmospheric emissions, which has become an essential tool for planning and implementing environmental and sustainable mobility policies.

It should be noted, however, that the COPERT model used by ISPRA to estimate road traffic pollution does not explicitly consider transit traffic at the regional level,

Table 3 GHG CO_2 eq emissions by region in Italy—value expressed in kton

Region	2021	%
Lombardy	62,860.418	20.4
Emilia Romagna	31,726.162	10.3
Apulia	30,708.317	10.0
Veneto	29,906.874	9.7
Lazio	26,551.477	8.6
Sicily	23,208.077	7.5
Piedmont	21,156.434	6.9
Tuscany	12,922.189	4.2
Campania	12,518.503	4.1
Sardinia	11,633.430	3.8
Friuli-Venezia Giulia	8,185.146	2.7
Calabria	6,718.167	2.2
Trentino South Tyrol	6,500.231	2.1
Liguria	5,693.798	1.9
Brands	5,308.296	1.7
Umbria	4,570.720	1.5
Abruzzo	3,444.989	1.1
Basilicata	2,078.645	0.7
Molise	1,784.771	0.6
Aosta Valley	282.396	0.1
Grand total	307,759.040	100

Source Our elaborations on ISPRA data 2021

focusing mainly on the assessment of emissions from internal road traffic within individual regions, taking into account the different local conditions and specific environmental needs.

With this study, an attempt is made to provide a dimension of the impact of pollutant emissions that is not exhausted within the individual region, such as that generated by the city traffic component, but by the flows of goods transport that originates within regions and is destined for foreign markets.

1.4 Freight Transport by Road for the International Market

Goods exports by road represent a key component of Italy's international trade, with a value that reached 12 billion euro in 2023, according to data provided by ISTAT in its annual report on foreign trade. This sector showed significant growth of 5% over the previous year, highlighting the resilience and competitiveness of Italian-made goods in foreign markets.

Table 4 GHG CO_2 eq emissions by road transport sector in Italy—value expressed in kton

Region	2021	%
Lombardy	16,371.642	16.95
Lazio	9,272.886	9.60
Emilia Romagna	9,106.393	9.43
Veneto	8,986.378	9.30
Piedmont	7,294.539	7.55
Campania	6,878.107	7.12
Tuscany	6,653.330	6.89
Sicily	5,864.231	6.07
Apulia	4,883.542	5.06
Brands	3,059.408	3.17
Trentino South Tyrol	3,031.255	3.14
Liguria	3,016.312	3.12
Calabria	2,368.970	2.45
Sardinia	2,321.236	2.40
Friuli-Venezia Giulia	2,168.628	2.24
Abruzzo	2,160.637	2.24
Umbria	1,598.710	1.65
Basilicata	680.577	0.70
Aosta Valley	469.806	0.49
Molise	419.874	0.43
Grand total	96,606.461	100

Source Our elaborations on ISPRA data 2021

Particularly noteworthy was the performance of exports to European Union countries, which grew by 7% in 2023. This figure underlines the importance of the European market for Italian road goods and our country's ability to meet the needs of a highly competitive and quality-conscious market.

Italian regions contributed differentially to these exports, with Lombardy leading the way with more than EUR 100 billion worth of goods exported, followed by Piedmont and Trentino Alto Adige. These figures highlight the importance of road freight transport in facilitating the movement of a wide range of products to international markets, underlining the crucial role this sector plays in the Italian economy and in global trade.

Looking at the volume of goods transported to foreign markets (Table 5).

Table 6 shows the volumes of goods traffic passing through the Alpine passes from Italy. The Felbertauern pass has the highest freight traffic volume with 5.4 ktonnes, followed by the Brenner pass with 5.3 ktonnes and the Reschen pass with 2.1 ktonnes.

Table 5 Volume of freight traffic exported from Italy to overseas destinations—value expressed in kton

Region	2021	%
Lombardy	4,306.857	29.5
Piedmont	2,538.974	17.4
Trentino South Tyrol	1,716.131	11.7
Veneto	1,582.402	10.8
Emilia-Romagna	1,351.944	9.3
Friuli-Venezia Giulia	628.044	4.3
Tuscany	533.903	3.7
Liguria	461.481	3.2
Campania	342.389	2.3
Apulia	332.617	2.3
Brands	217.858	1.5
Lazio	159.001	1.1
Aosta Valley	138.498	0.9
Sicily	98.263	0.7
Abruzzo	90.862	0.6
Umbria	32.089	0.2
Basilicata	30.227	0.2
Calabria	20.371	0.1
Sardinia	18.103	0.1
Molise	12.711	0.1
Grand total	14,612.725	100.0

Source Our elaborations on ISTAT data 2021

Table 6 Volume of freight transport through major Alpine passes—value expressed in tonne

Code_value	Crossing	Road (ton)	Railway (ton)	Total (tonnes)
1087	Montgenevre	651.800	–	651.800
1291	Moncenisio/Frèjus	11,598.300	2,863.700	14,462.000
7022	Mont Blanc	9,544.700	–	9,544.700
7064	Great St. Bernard	2,079.700	–	2,079.700
8065	Ventimiglia	20,961.400	743.600	21,705.000
13110	Reschen	8,826.500	15,110.800	23,937.300
21010	Brenner	39,918.900	13,781.900	53,700.800
30117	Felbertauern	61,017.800	23,832.600	84,850.400
103067	Simplon	1,078.100	11,538.200	12,616.300

Source Our elaborations on Alpine Traffic Observatory European Commission 2021

1.5 Impact of New Technologies on Italian Exports

As mentioned in the previous section, road goods exports represent a key component of Italy's international trade with growth levels of around 7% to EU countries. A key factor in the growth of road exports has been the impulse provided by digital exports. According to data provided by the Osservatorio Export Digitale (Digital Export Observatory) of the School of Management of the Politecnico di Milano, B2B digital exports reached a value of EUR 175 billion in 2022, registering a growth of + 20% compared to the previous year. This sector accounts for around 28% of total Italian exports, confirming the importance and potential of digital exports for Italian companies.

In the most digitized sectors, such as Automotive, Fashion, and Mechanics, significant increases have been recorded, with the pharmaceutical sector standing out with a growth of + 47%. These data highlight how the adoption of digital solutions is contributing substantially to the growth and competitiveness of Italian exports, enabling companies to expand their presence in international markets and improve the efficiency of business operations.

As far as B2C digital exports are concerned, the leading sectors are Fashion, Food and Beverage and Furniture. In particular, the Fashion sector recorded a value of EUR 10.1 billion, representing 54% of total B2C digital exports. These figures confirm the importance of digital in favoring the diversification of Italian exports and promoting the growth of key sectors of the national economy.

Against this background, current trends indicate that the volumes of road freight transported to foreign countries could continue to grow significantly in the coming years. According to a study conducted by the Italian Logistics and Transport Association, there will be a 6% annual increase in the volume of goods transported by road to the main foreign markets in the period 2020–2023, mainly driven by the growing demand for high-quality Italian products and the increased logistics efficiency in the road transport sector.

This positive trend has been supported by the increased demand for Italian goods in foreign markets and the efficiency of road transport services, which have made this sector increasingly competitive internationally.

The introduction of new technologies in the transport sector has further contributed to boosting road freight exports. The adoption of advanced tracking systems based on the Internet of Things (IoT) has resulted in a 15% reduction in delivery times and a 20% decrease in operating costs for road exports. Furthermore, the use of route optimization algorithms and real-time goods flow management systems has improved supply chain management and delivery planning, enabling Italian companies to offer faster and more personalized services to foreign customers.

Finally, the adoption of artificial intelligence (AI) technologies is helping to improve road freight management, enabling real-time data analysis and more informed decision-making. AI can be used to optimize delivery planning, predict delivery times and manage exceptions, improving road freight management and reducing delivery times.

Looking to the future, industry projections indicate that the volumes of road freight transported to foreign countries could continue to grow, with a projected increase of 5% per year over the next three years.

A specific scenario simulation will be performed on this value for the CO_2 impact assessment by applying the analysis model proposed by the authors.

2 A Model for Environmental Impact Assessment

In order to assess the impact of CO_2 emissions generated by road freight transport, an analysis model using specific data on freight volumes destined for foreign markets and a calculation of CO_2 emissions based on road distance is proposed below.

2.1 Transport Model for the Allocation of Goods

The transport supply system adopted in this work is of the mono-modal type, assuming that goods move through space using private vehicles. The transport service will therefore be continuous and simultaneous as it is available and accessible throughout the network.

The cost of road transport is defined, in a simplified manner, assuming that it is a function of a single explanatory variable, namely the kilometer distance between the place of production and the location of the Alpine passes; more specifically, the cost of crossing the generic arc i is:

$$\mathrm{T_i} = \mathrm{t_i} \times \mathrm{l_i}$$

where $\mathrm{l_i}$ represents the relative length of the individual arc, $\mathrm{t_i}$ is a parameter expressing the distance in kilometers. It is clear that the time cost of a generic trip k will be the result of the sum of the individual time costs of the component arcs of the graph.

The simulation of transport flows, resulting from the freight allocation process, is based on a gravity model modified in such a way that the supply is proportional to the access capacity of individual crossings.

In particular, with regard to the allocation of goods flows to transalpine crossings, which in terms of cost represent the best solution, a transport model was used, which in its standard theoretical formulation for this type of problem is represented by the following objective function:

$$minimize\ Z = \sum_{i=1}^{m} \sum_{j=1}^{n} c_{ij} x_{ij}$$

$$subject\ to: \sum_{j=1}^{n} x_{ij} = D_i \quad \forall i = 1, \ldots, n$$

$$\sum_{i=1}^{m} x_{ij} = S_j \quad \forall j = 1, \ldots, m$$

$$x_{ij} \geq 0 \quad \forall ij$$

where Z represents the total cost of moving from municipality i to pass j. D_i identifies the export demand originating from municipality i (1, …, m), S_j indicates the access capacity j (1, …, n) at the transalpine crossings.

Since precise data on the geographic origin of goods is not available, a unit proportional to the presence of enterprises and number of employees located in Italian municipalities was assumed. If, in reality, differences should emerge in the requirements calculated in this way, it is possible to attribute these differences to intrinsic qualitative characteristics of the territory and the production fabric, which condition demand and, therefore, need to be studied.

2.2 *Simulation of Transport Flows*

The simulation of road freight flows originating from Italian municipalities and directed towards Alpine passes was generated using an origin–destination matrix, which provided a comprehensive representation of freight movements between different regions of the country. The transport flow assignment model was based on gravitational coefficients, which took into account the distance between origins and destinations, as well as the loading capacity of vehicles and the available transport infrastructure.

Simulation results showed that road freight traffic flows are mainly concentrated on the largest and busiest Alpine crossings, such as the Reschen pass, which recorded the highest freight traffic volume, followed by the Felbertauern and Brenner, as shown in Table 7.

CO_2 emissions were calculated by applying the following formula:

$$CO_2\ \text{emissions (t)} = \text{Volume transported (t)} \times \text{total distance (km)} \times \text{emission factor (t/t km)}$$

where, the transported volumes are represented by the total weight of goods expressed in tonnes (t); the total distance is calculated as the kilometric distance travelled from the node of origin, in this study represented by the 7785 Italian municipalities, to reach as destination the 9 transalpine passes; the emission factor (in kg CO_2/t km) is a coefficient representing the CO_2 emissions per unit of tonne-kilometer transported. This factor varies depending on the mode of transport used (road, rail, ship, air), vehicle/

Table 7 Simulated volume of freight transport through major Alpine passes by road—value expressed in ton

Crossing	Road (ton)	%
Reschen	4,657,226.22	31.9
Felbertauern	3,072,960.61	21.0
Brenner	3,072,688.55	21.0
Moncenisio/Frèjus	1,532,361.19	10.5
Ventimiglia	891,360.20	6.1
Simplon	708,801.54	4.9
Mont Blanc	497,395.73	3.4
Great St. Bernard	136,257.21	0.9
Montgenevre	40,532.34	0.3
Total	14,609,583.60	100

vehicle characteristics and operating conditions. Some typical emission factor values are 60–150 g CO_2/t km (road transport), 15–60 g CO_2/t km (rail transport), 10–40 g CO_2/t km (maritime transport), 500–1000 g CO_2/t km (air transport).

In the present study, assuming that goods are only transported by truck, it was decided to use an emission factor of 100 g CO_2/t km in the formula.

For the calculation of pollutant emissions, simulations were conducted using two scenarios:

- Scenario 1: Simulation with freight volumes transported by road in the year 2021[a].
- Scenario 2: Simulation with volumes of goods transported by road in the year 2021 increased annually by 5% until 2024[b].
- Scenario 3: Simulation with volumes of goods transported by road in the year 2021 increased annually by 5% until 2024 and simultaneous closure of the Brenner Pass[c].

Below are the CO_2 emission values simulated with road transport flows using the calculation formula described above for the first scenario (Table 8).

Comparing the data collected by ISPRA with the values resulting from the simulation, the distribution of CO_2 emissions among the regions maintains roughly the same proportions, suggesting a good level of calibration of the transport model by macro area. More significant differences between actual and simulated values can be observed in the northern regions, where, compared to the data collected by ISPRA, the simulation of transport flows returns higher emission levels in Piedmont and Veneto for the northern regions, Apulia and Campania for the southern regions while lower emission levels in Lazio for the central regions (Fig. 3).

We then proceeded with the simulations of the second scenario, for which, starting from the input data of the first scenario, we assumed an increase in volume every year of 5% up to 2024 while keeping the foreign market access crossings unchanged, and of the third scenario, for which, in addition to the 5% annual increase in volume, we also simulated the closure of one of the most important international crossings for access to foreign markets: the Brenner Pass (Table 9).

Table 8 Simulated CO_2 emission by road freight transport through major Alpine passes 2021—value expressed in tonnes

Region	2021[a]	%
Lombardy	61,207	14.30
Emilia-Romagna	53,324	12.46
Veneto	47,274	11.05
Piedmont	46,878	10.95
Apulia	39,209	9.16
Campania	34,975	8.17
Trentino-Alto Adige	30,964	7.24
Tuscany	26,690	6.24
Sicily	16,995	3.97
Brands	14,149	3.31
Lazio	12,713	2.97
Liguria	12,326	2.88
Friuli-Venezia Giulia	11,058	2.58
Abruzzo	7,408	1.73
Basilicata	3,491	0.82
Calabria	2,881	0.67
Umbria	2,130	0.50
Sardinia	1,597	0.37
Aosta Valley	1,447	0.34
Molise	1,201	0.28
Grand total	427,917	100

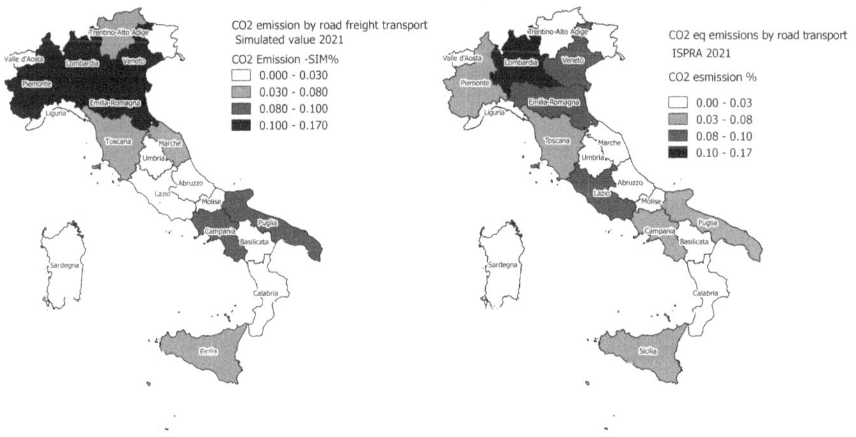

Fig. 3 CO_2 eq emission by road freight transport—ISPRA 2021 versus simulation

Table 9 Simulated CO_2 emission by road freight transport through major Alpine passes—value expressed in ton

Region	2021[a]	2024[b]	Δ% 2021[a]–2024[b]	2024[c]	Δ% 2021[a]–2024[c]
Lombardy	61.207	70.388	15	74.030	21.0
Emilia-Romagna	53.324	61.322	15	68.641	28.7
Veneto	47.274	54.365	15	62.477	32.2
Piedmont	46.878	53.910	15	55.024	17.4
Apulia	39.209	45.090	15	50.770	29.5
Campania	34.975	40.221	15	45.175	29.2
Trentino-Alto Adige	30.964	35.609	15	80.634	160.4
Tuscany	26.690	30.693	15	33.929	27.1
Sicily	16.995	19.544	15	21.979	29.3
Brands	14.149	16.272	15	18.292	29.3
Lazio	12.713	14.620	15	16.387	28.9
Liguria	12.326	14.175	15	14.679	19.1
Friuli-Venezia Giulia	11.058	12.717	15	13.097	18.4
Abruzzo	7.408	8.519	15	9.580	29.3
Basilicata	3.491	4.014	15	4.510	29.2
Calabria	2.881	3.313	15	3.724	29.3
Umbria	2.130	2.449	15	2.741	28.7
Sardinia	1.597	1.837	15	2.034	27.4
Aosta Valley	1.447	1.664	15	1.688	16.7
Molise	1.201	1.381	15	1.551	29.1
Total	427.917	492.105	15	580.941	35.8

Third column (2024[b]) where the results of the second simulation are reported, which includes a 15% increase in volumes compared to the first simulation. Fourth column (Δ% 2021-2024[b]) indicating the percentage change between the emissions of 2021 and those generated by the second simulation. Fifth column (2024[c]) where the results of the third simulation are reported, which includes both a 15% increase in volumes and the closure of the Brenner Pass. Sixth column (Δ% 2021-2024[c]) indicating the percentage change between the emissions of 2021 and those generated by the third simulation

The results of the simulations show that, with an increase in the number of tonnes transported by road, the resulting increase in CO_2 emissions is proportional to the volume loaded, i.e. + 15%, and the contribution of the individual regions to the production of this pollutant is unchanged. Different results are obtained with the third scenario, for which, in addition to an increase in tonnages handled, the Brenner Pass, one of the nine modeled crossings, is to be closed. In this case, the increase in CO_2 emissions is + 35.8% compared to 2021, with the deviation influenced by the redirection of transport flows from all regions. Particularly noticeable is the 160% change recorded in Trentino-Alto-Adige, the third largest region in terms of exported volumes (ISTAT 2021), for which the closure of the Brenner Pass, considered the preferred route for exports and transport of goods, would have a considerable impact on the redirection of transport volumes and consequently on CO_2 emissions into the atmosphere.

3 Conclusions

The high demand for road transport of people and goods has led to a significant impact on air quality and climate change in all developed regions. According to the European Environment Agency, the transport sector is responsible for about 25% of total greenhouse gas emissions in the European Union, of which 72% is attributed to road transport.

In order to assess the impact of CO_2 emissions generated by road freight transport, an analysis model is proposed that uses specific data on freight volumes destined for foreign markets and a calculation of CO_2 emissions based on road distance. The simulation of transport flows is built on an origin–destination (O/D) matrix and calculates CO_2 emissions based on the tonnes of goods transported and the distance travelled to reach Alpine crossings.

In order to calibrate the model, foreign transport demand data equivalent to those recorded in the year 2021 were used.

The analysis conducted on the CO_2 emissions generated by road freight transport to foreign markets shows a diversified impact by macro-area of the territory where the northern regions, with a higher density of industrial and commercial activities, contribute significantly to CO_2 emissions. Also of non-negligible importance is the contribution of some regions of central and southern Italy, which are characterized by a lower production capacity compared to the northern regions, but which nevertheless, given the greater physical distance in reaching foreign markets, generate emissions with a significant impact on the environment.

It should be pointed out that the proposed study is based on a single-mode transport model and that a natural consequence of this is to carry out simulations on a multimodal transport model by integrating two O/D matrices, one for road transport and one for rail transport, in order to analyze possible reductions in the impact of CO_2 emissions as well as consequent savings for the production system.

Therefore, a possible evolution of the model by integrating different modes of transport would lead to the emergence of alternative solutions capable of mitigating the emissions generated by companies in Southern Italy, such as a greater use of intermodal terminals, such as those located in Calabria, Campania and Lazio (Fig. 4), with the transfer of part of the road transport to rail transport. These terminals in Italy play a crucial role in the country's freight transport system as they facilitate the seamless transfer of goods between different modes of transport, such as rail, road and sometimes sea or air.

In conclusion, this study provides a methodology for analyzing the environmental impact of road freight transport, with a specific focus on CO_2 emissions and routes

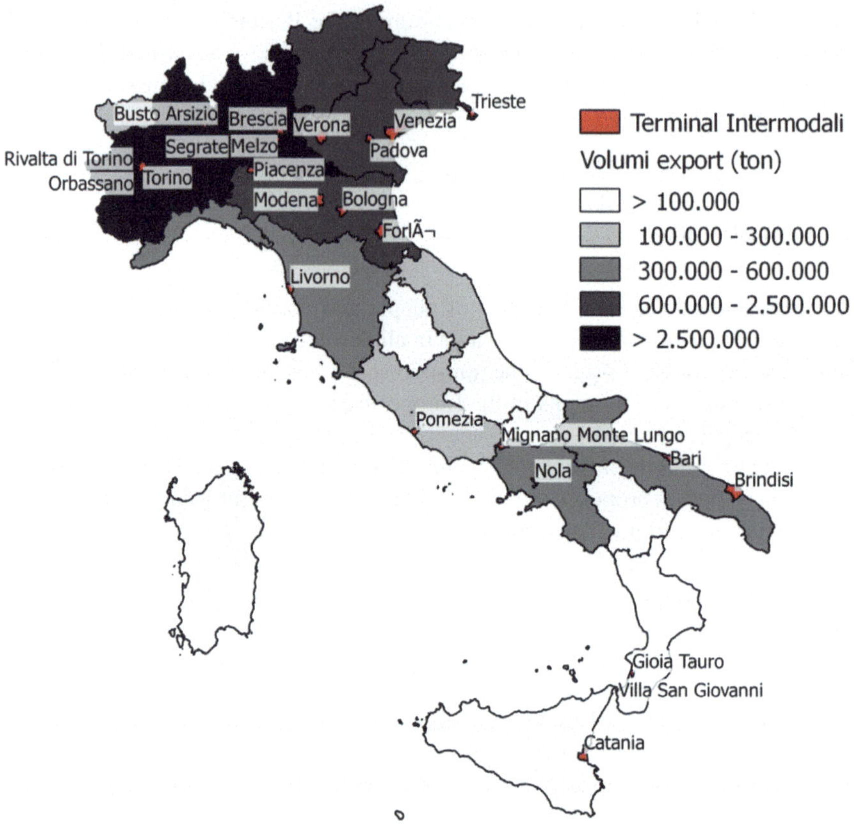

Fig. 4 Intermodal terminals in Italy for rail freight transport

crossing Alpine passes. The results can be used to support the transition to a more sustainable and environmentally friendly transport system that is essential to address the significant contribution of this sector to global greenhouse gas emissions, in particular carbon dioxide.

Furthermore, for Italian companies wishing to internationalize, an efficient and sustainable logistics model is crucial to ensure long-term competitiveness and sustainability. An intermodal logistics model that includes rail and maritime transport in addition to road transport can help reduce the overall environmental impact of transport, improve the efficiency of goods flows and reduce transport costs, and contribute to a more sustainable and environmentally friendly future.

References

1. Green KW, Zelbst PJ, Meacham J, Bhadauria (2012) Green supply chain management practices: impact on performance. Supply Chain Manage 17(3):290–305
2. Peng Z, Wu Q, Li M (2020) Spatial characteristics and influencing factors of carbon emissions from energy consumption in China's transport sector: an empirical analysis based on provincial panel data. Pol J Environ Stud 29(1):217–232
3. Benedetti E, Panzone L, Wildman J, Seal C (2023) The effect of carbon tariffs on UK food trade and its carbon emissions: a gravity model with micro-simulation. SSRN Electron J 20
4. Xiang W, Xiaoyan C, Libang M, Jing B, Man L, Simin Y (2023) Global and regional trends in greenhouse gas emissions from rice production, trade, and consumption. Environ Impact Assess Rev 101
5. Zhao X, Wise MA, Waldhoff ST, Page Kyle G, Huster JE, Ramig CW, Rafelski LE, Patel PL, Calvin KV (2022) The impact of agricultural trade approaches on global economic modeling. Glob Environ Change 73
6. Randrianarisoa LM, Gillen D (2022) Reducing emissions in international transport: a supply chain perspective. Transp Res Part D Transp Environ 102
7. Green Jr KW, Zelbst PJ, Meacham J, Bhadauria VS (2012) Green supply chain management practices: impact on performance. Supply Chain Manage Int J 17(3):290–305
8. Murphy PR, Poist RF (2003) Green perspectives and practices: a "comparative logistics" study. Supply Chain Manage Int J 8(2):122–131
9. Singh RK, Murty HR, Gupta SK, Dikshit AK (2009) An overview of sustainability assessment methodologies. Ecol Indic 9(2):189–212
10. Linton C, Grant-Muller S, Gale WF (2015) Approaches and techniques for modelling CO_2 emissions from road transport. Transp Rev 35(4):533–553
11. Chen J, Xu C, Li K, Song M (2018) A gravity model and exploratory spatial data analysis of prefecture-scale pollutant and CO_2 emissions in China. Ecol Ind 90:554–563
12. Faruque MO, Rabby MAJ, Hossain MA, Islam MR, Ur Rashid MM, Muyeen SM (2022) A comparative analysis to forecast carbon dioxide emissions. Energy Rep 8:8046–8060
13. ISPRA homepage. https://www.isprambiente.gov.it/it. Accessed 2024/1/4
14. ASVIS homepage. https://asvis.it/. Accessed 2024/1/4
15. EEA homepage. https://www.eea.europa.eu/en. Accessed 2024/1/4
16. COPERNICUS homepage. https://www.copernicus.eu/en. Accessed 2024/1/4
17. EDGAR homepage. https://edgar.jrc.ec.europa.eu/. Accessed 2024/1/4
18. Ecotransit homepage. http://www.ecotransit.org/. Accessed 2024/1/4
19. ISTAT homepage. https://www.istat.it/it/archivio/13955. Accessed 2024/1/4

Artificial Intelligence, Territory, Town, and Anthropization Processes

Stefano Aragona

Abstract Artificial Intelligence, AI, can be of great use in the anthropization processes of the territory and urban centers. For many years, territorial planners and urban planners have been calling for the creation of Observatories to follow, evaluate choices and implementations. What Kevin Roose reports in the 2023 article in the New York Times about conversations with AI devices is astonishing [18]. Although several of the chatbot's statements and responses present some inconsistencies and could be a free association—but with internal coherence—of the billions of data, of a multidisciplinary nature. Of this we have no proof of the opposite nor of the correctness of this sentence except through "human" verification. With AI there is the possibility of making a sort of increasingly larger virtual reproduction of what is observed. That is, the holistic approach can be increasingly approached. And this allows the construction of increasingly complex and "complete" scenarios. But precisely because of the methods of building AI, i.e. feeding it with "data" which are in any case "phenomenal" even if translated into quantitative and/or qualitative indicators, it is still a "partial" action both from a territorial perspective both environmental and social. Action which in fact, in relation to these aspects, risks being "conservative" of the status quo. Therefore, the writing highlights these aspects, potential and limitations in order to provide the most operational indications possible.

Keywords Anthropization processes · Town Planning · Artificial Intelligence

1 Space and Choices of Innovation

Since when they are—in theirs various evolutions, to the current state of knowledge that began 7 million years ago with *Sahelanthropus tchadensis* discovered in Chad, before the separation of the evolutionary line of man which occurred 1 million years later—the first hominids who probably could had walk having two legs [17]. With

---

S. Aragona (✉)
National Institute of Bioarchitecture—INBAR Rome Section, Rome, Italy
e-mail: stefano.aragona@gmail.com

D. Marino and M. Monaca (eds.), *Generative Artificial Intelligence and Fifth Industrial Revolution*, Lecture Notes in Networks and Systems 880,
https://doi.org/10.1007/978-3-031-73880-7_13

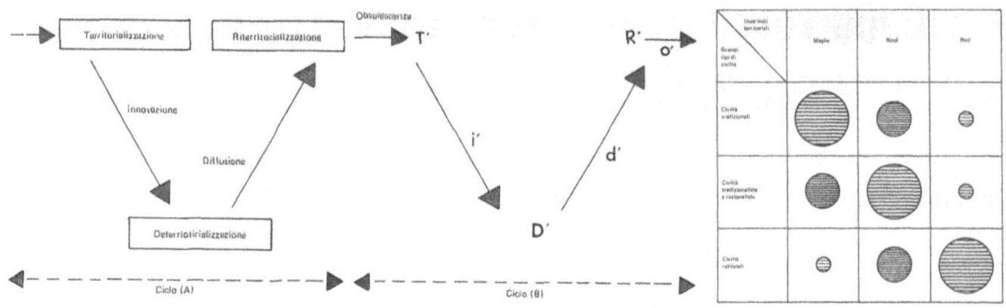

Fig. 1 Territorialization, deterritorialization, reterritorialization, innovation, and territorial invariants [16]

the possibility of walking there is the first opportunity to modify, temporarily or permanently, the space even if it is just crossed. And much later, 6 million years pass, there are the first stable human settlements that we know today: they are in Anatolia and date back to around 12,000 years ago [22].

All the steps and changes are linked to technological innovation. Thanks to this, Raffenstein [16] theorizes, there are processes of *territorialization, deterritorialization, reteritorialization.* These dynamics modify the relative weight of what he identifies as *Territorial Invariants*—that is, the *areas*, *nodes*, and *networks*—that characterize the various forms of cultures and societies (Fig. 1).

The initial purpose of all anthropic transformations is to improve living conditions. But this aim is increasingly starting to be accompanied by that of increasing value. In modern terms it can refer to the product and/or service. Regarding this second element, with the innovations linked to advanced telecommunications, which later became telematics, the concept of *Value Added Service* (VAS) was born. This implies a growth in *exchange value* to the detriment of *use value.*

Therefore, if there is widespread information and sharing of choices, it is possible to redistribute the income deriving from the VAS. This situation can increase workers' disposable income, improve working conditions and participate in environmentally and socially sustainable development. In this regard, in 2020 Logroscino cites the example of Finland where, also thanks to digital innovation, working hours are being reduced for the same salary: a concrete example that a "metaverse" that helps social and economic equity is possible [9].

Conversely, if AI is in the hands of a few political, or worse economic, decision makers as seems the things are going on, in the physical world, in the social and economic dimensions, it will create "mass control" and an unfree society as Huxley [6, 7] and Orwell [13] wrote in their famous texts that they seemed like science fiction.

AI certainly has the possibility of artistic creativity but regarding social creativity it, in its current state, cannot make sense. There are many examples of works of art resulting from the mix of the multitude of information provided to AI systems. Thanks to this knowledge it then created "products" that have quotes deriving from everything that was provided to it. It is a creative mechanism that seems very similar to what human beings have, more or less consciously. At the moment we cannot know whether the AI carries out this creative action with the awareness of the choice or following connections linked to the amount of information loaded or to a path of choice that it decides to have. Probably this aspect will be able to be investigated the more neuroscience studies and knowledge continue.

Instead, in relation to social creativity or actions linked to actions of association, proposal, protest and social struggles and so on, AI is not a conscious actor but "only" coherent or not coherent with the ethical and ideal references that have been introduced in his "brain".

And this aspect is truly problematic since these references are not neutral but represent values of each culture, of each legal and institutional system. This means that AI is "conservative" of different and varied contextual Systems. But in this way there is no possibility of "ruptures in the System" and also of the introduction of "diversity" into the various societies. Therefore the comparison between "different", the basis of creativity as highlighted for years [4], becomes practically impossible. And the result of this situation is the progressive consolidation of the social and, consequently, also cultural and even scientific and technological status quo. If we were to make a biological metaphor it is as if in a community there were marriages only between relatives: which significantly leads to madness as numerous scientific research shows.

The overall consequence of being "conservative" would lead to an increasing uniformity with a decrease in resilience, firstly social and subsequently spatial and environmental: in fact, AI would act according to efficiency criteria, a sort of "Paretano Optimum", but not social effectiveness if not that which leads to undifferentiated homogeneity. In short, the outcome would be the standardization of human beings and places that would lose their specificity and become just spaces.

2 AI Increases Territorial Inequality if There Is No Alliance Between Innovation and Places

Regarding AI as a tool for dealing with various territorial issues, there is the risk of a serious availability gap between the main urban centers and the so-called internal or "minor" areas. This is because AI works on databases provided by many and multiple institutional, formal and non-formal sources [11]. These are focused on the major economic and social attractors while neglecting and paying much less attention to the rest of the territory. So if AI is guided by a thought of economic efficiency, the differences between territories will increase.

It creates locational advantages and disadvantages. This was already highlighted by Gillespie and Robin in 1989 with the development, and then diffusion, of advanced, increasingly interactive telecommunications, later called *telematics* [1, 5]. In more recent times this is named *digital devide* [12]. Therefore, the more in a

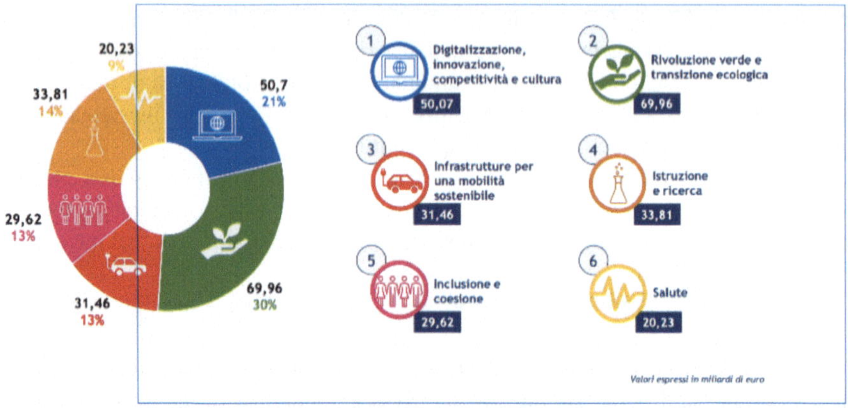

Fig. 2 Resource allocation in the PNRR [15]

territory there is the possibility of connections that allow remote, or telematic, activities, the more attractive and therefore competitive it can be. This aspect is one of the key elements in the *National Strategy of Internal Areas*, SNAI, dedicated to reducing the delay of these territories with respect to the rest of the country. Such a fundamental aspect that *digitalization* is one of the pillars of the *National Recovery and Resilience Plans*, PNRR: in the Italian one it is a quarter of the total huge financial allocation (Fig. 2).

At present AI is promoted and supported by economic actors who, legitimately, have profit as their aim. It requires significant infrastructure and therefore huge costs. This means that it risks increasing territorial inequalities, a risk for internal areas of being even more peripheral. Romiti in 2022 notes that the Port cities as new digital hubs for technological reasons connected to the (international) sub-marine cables capacity underlying that 97% of internet traffic travels through submarine cables [19]. In December 2021, Europe launched the "EU Global Gateway" initiative, a 300 billion investment to support the connection between member countries and the rest of the world. The *Internet of Things* has reached 12.3 billion connections, there are 4.9 billion people with internet access (2021 edition of Facts and Figures, the annual report of the International Telecommunication Union), almost 60% live in urban contexts, in Italy this reaches 72%.

There are numerous investments for the reduction of the digital divide that the Italian State has implemented:

- The National Strategy for Ultra-Broadband "Towards the Gigabit Society" aims to reduce the digital divide in rural areas and bring connectivity to 1 Gbps throughout Italy by 2026.
- The digital cohesion of the internal areas is included in the package of the PNRR:
  - Mission 1, *Digitalization, Innovation, Competitiveness.*
  - Mission 5 *Inclusion and Cohesion*, as well as in the implementation.

- Interventions of the *Cohesion Policy of the European Structural and Investment Funds 2021–2027*.

But they are probably not effective against what is called the "new internal area divide". The reason derives from the basic elements of AI systems that is:

- data volume
- high-performance computers
- algorithmic operations.

Today the internal areas are still poorly "sensorized":

- the endogenous data production is more modest
- the volume of data, structured or not, generated by heterogeneous sources
- variety and speed are limited.

The machine learning descended from these input factors determines a reliability and a value of the output that is difficult to accept as conventional. Therefore:

- the ability of the algorithm to perceive the context
- learn from the data.

To reason on a generally acceptable final information is precluded by the initial conditions.

This is where a third way of developing artificial intelligence comes forward, proclaimed by Craig Dalton and Jim Thatcher [2], two geographers, promoters of a "*... policy that reinterprets data and algorithms as tools aimed at emancipation and solidarity*".

Developing the technology stack of AI systems for internal areas requires a holistic view and public accountability 3.0. Co-design and co-evolution of these systems should enter the Internal Areas National Strategy—SNAI [10] by right. Starting from this planning period, we can begin with the construction of new ex-ante detection indicators—volume of data produced and number of homogeneous sources—and the attribution of service outcomes released by dynamic algorithmic information.

This path can be followed and facilitated if accompanied by a process of territorial reorganization of services and by supra-administrative governance involving local authorities in a logic of sharing a single data space, originating in all internal and extra-area areas, and adherence to a taxonomy of interdisciplinary learning modalities.

3 Some Emerging Ideas for Temporary Conclusions in a Missing Integrated Territorial Planning

With Meloni Government there is a lack of a Territorial vision of Planning on a national scale. The statewide SNAI is downplayed and the choices are essentially entrusted to the Regions. The Internal Areas theme no longer appears on the Government Agenda, even more so if associated with the "Southern question".

The issues of territorial rebalancing linked to national and regional choices are not in the Government's political vision. And what is being done often finds dissent of the territorial actors see e.g. the inadequate the Call for small municipalities of 2023.07.14, law n. 158/2017 [21].

And there is a question of words and political choices. The Meloni Government has "banned" some keywords, basic concepts, linked to the UN 2030 Agenda for Sustainable Development, among the main ones:

- Ecological transition and therefore Renewable Energies, with the paradox of proposing Italy as the European Gas HUB
- Circular Economy
- Ecological Footprint
- 0 Soil Consumption
- Fair and Equitable Wellbeing (BES, ISTAT—CNEL, since 2013).

While digitization is seen only as a potential increase in efficiency especially regarding the Public Administration, not considering its great potential in terms of social effectiveness and for more equitable and balanced territorial arrangements as required since 2007 in the UE Leipzig Charter, then updated in 2020 by the European Commission [3, 20].

About renewable energies Meloni Government, following and expanding the choices of the Draghi Government, cancels all the Energy and Seismic Security bonuses of the Conte 2 Government that is:

- It blocks the safety of the existing buildings, at 75% outdated
- It blocks the possibility of "on site" and increasingly autonomous energy production
- It sinks the micro-entrepreneurship that was being created by combining:
  - Requalification of the building fabric tissue
  - Economic activities (some estimates speak of around 125,000 small businesses and 500,000 employees involved).

The economic vision is the liberistic one based on microeconomic criteria and not inspirated to macroeconomics which instead considers the Keynesian economic multiplier.

Using the AI in this general context of political choices can lead to the vision of *The New Town* by Sant'Elia: No Public Spaces, Only Individuals, Disappearing of Society i.e. Disappearing of the Town. But with the constrain of Metropolis by Fritz Lang and as "outcome" the Ridley Scott's Blade Runner City (Fig. 3).

While it, even before the pandemic due to COVID, planning experiences were being experimented on a territorial scale to combine "green" and "digitization" at a territorial scale, i.e. the most suitable one to move towards ecological and socially and environmentally sustainable development. The Municipality of Pegognaga, province of Mantua, in 2017 had started a "Smart Land" path (Fig. 4) thus anticipating the proposal of what in the PNRR will be the *Ecological Transition*.

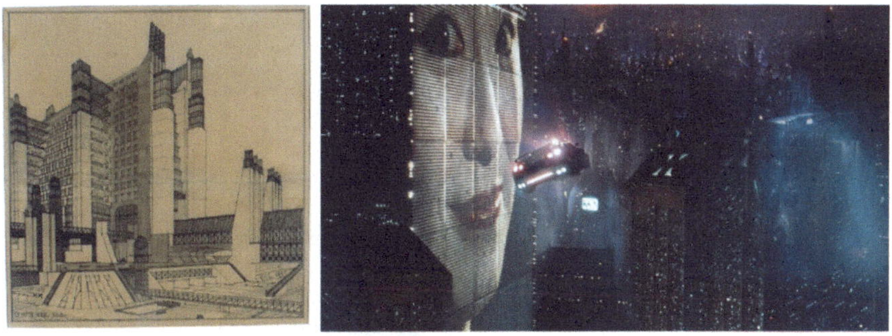

Fig. 3 The new town by Sant'Elia, left, and in Blade Runner, right (Wikipedia Commons, Jon Fusco, 2017)

Fig. 4 The Smart Land proposed in 2017 by Pegognaga (MN) (Municipality of Pegognaga (MN)) [14]

And in all of this, AI is a very powerful tool if you wanted to educate it, that is, nourish it, in this sense. That is, considering the territory and urban centres, i.e. anthropizations as a common good to be thought about, planned, designed and implemented together with local communities, following the philosophy of the Encyclical *Laudato Sii for the Care of the Common Home* [8].

We are now all aware that the ecological footprint of everything we build, and of its management (Fig. 5), is an insurmountable limit already indicated in the 1972 Report in *The Limits to Growth*: basic elements of the United Nations 2030 Agenda for Sustainable Development. Thus AI could offer us enormous and unexpected opportunities of the circular economy, that is, the cornerstone of new methods of development and therefore of anthropization, allowing us to ride on the long wave of the Ecological Transition.

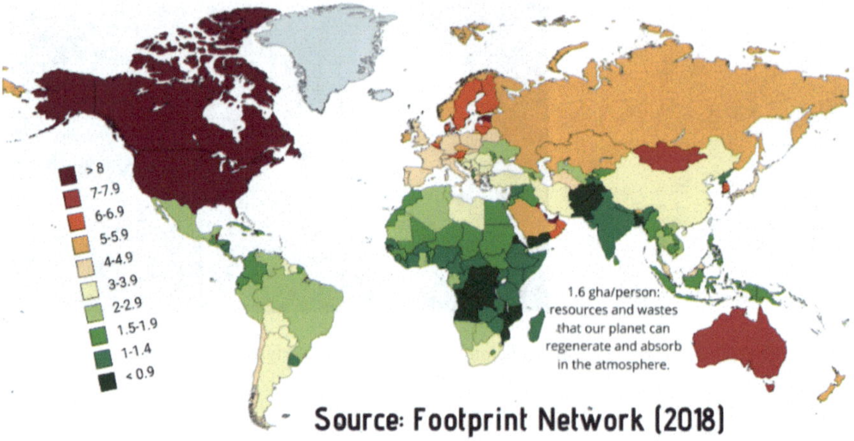

Fig. 5 Ecological footprint 2018 (Footprint Network, 2018)

References

1. Aragona S (1993) La città virtuale. Trasformazioni urbane e nuove tecnologie dell'informazione. Gangemi, Roma—Reggio Calabria
2. Dalton GM, Thatcher JE (2021) Data power: radical geographies of control and resistance. Pluto Press, London
3. EU - European Commission (2020) New Leipzig Charter—the transformative power of cities for the common good. https://ec.europa.eu/regional_policy/en/newsroom/news/2020/12/12-08-2020-new-leipzig-charter-the-transformative-power-of-cities-for-the-common-good. Adopted at the Informal Ministerial Meeting on Urban Matters on 30 November 2020. www.bmi.bund.de. Accessed 28 Dec 2023
4. Florida R (2003) L' ascesa della nuova classe creativa. Stile di vita, valori e professioni. Mondadori, Milano
5. Gillespie AE, Robin K (1987) Geographical inequalities: the spatial bias of the new communications technologies. J Commun
6. Huxley A (1932) Brave new world, Ist edn. Mondo Nuovo. Medusa, Arnoldo Mondadori
7. Huxley A (1961) Brave new world revisited, Ist edn. Ritorno al Mondo Nuovo. Arnoldo Mondadori Editore, Milano
8. Lettera Enciclica Laudato Sii del Santo Padre Francesco sulla Cura della Casa Comune (2015) Tipografia Vaticana, Città del Vaticano
9. Logroscino A. Lavorare meno (lavorare tutti) in tempi di covid, la finlandia rilancia. Premier vuole giornata di 6 ore: 'Saremo più equi e produttivi'. https://www.ansa.it/canale_lifestyle/notizie/societa_diritti/2020/08/25/lavorare-meno-lavorare-tutti-in-tempi-di-covid-la-finlandia-rilancia_465a6725-bc4d-45d4-bc54-89c6e4fad4e9.html. Accessed 09 Jan 2024
10. Lucatelli S (2015) La strategia nazionale, il riconoscimento delle aree interne. Franco Angeli, Milano
11. Marino D, Gil Lafuente J, Tebala D (2023) Innovations and development of artificial intelligence in Europe: some empirical evidences. Eur J Manag Bus Econ 32(5):620–636. https://doi.org/10.1108/EJMBE-03-2023-0085
12. Micheli M (2022) Tutti i divari digitali dell'Italia: ecco perché la tecnologia crea nuove disuguaglianze, 19 Apr 2022. https://www.agendadigitale.eu/smart-city/ia-e-aree-interne-si-rischiano-nuove-periferie-cosa-deve-fare-la-politica-di-coesione/. Accessed 31 Jan 2024
13. Orwell G (1950) Nineteen eighty-four (1949), 1984. Oscar Mondadori, Milano

14. Pegognaga, Comune (MN): Smart Land Centro Culturale Livia Bottardi Milani (centroculturalepegognaga.it). http://www.centroculturalepegognaga.it/smart-land/. Accessed 15 Jan 2024
15. Presidenza del Consiglio dei Ministri (2021) Piano Nazionale di Ripresa e Resilienza. #NEXTGENERATIONITALIA
16. Raffestin C (1984) Territorializzazione, deterritorializzazione, riterritorializzazione e informazione. In: Turco A (ed) Regione e regionalizzazione. Angeli, Milano
17. Redazionale Focus. Scienze La timeline dell'evoluzione dell'uomo. Breve storia del genere Homo. https://www.focus.it/scienza/scienze/la-timeline-dellevoluzione-delluomo. Accessed 15 Jan 2024
18. Roose K. GPT-4 Is Exciting and Scary. Today, the new language model from OpenAI may not seem all that dangerous. But the worst risks are the ones we cannot anticipate, The New York Times, https://www.nytimes.com/2023/03/15/technology/gpt-4-artificial-intelligence-openai.html
19. Romiti S (2022) DIGITALE E DISUGUAGLIANZE. IA e aree interne, si rischiano nuove periferie: cosa deve fare la politica di coesione, 3 Nov 2022. https://www.agendadigitale.eu/smart-city/ia-e-aree-interne-si-rischiano-nuove-periferie-cosa-deve-fare-la-politica-di-coesione/. Accessed 31 Jan 2024
20. UE (2007) Leipzig Chart on the European sustainable cities. http://www.sinanet.isprambiente.it/gelso/files/leipzig-charter-it.pdf
21. UNCEM (2023) Tutti contro tutti Bandi INADEGUATI Marco Bussone Presidente, 14 July 2023. Uncem uncem.nazionale@uncem.net tramite antherica.com
22. Wienke U (2013) Storia dell'architettura climatica. https://architetturaclimatica.blogspot.com/2013/09/i-primi-insediamenti-umani-stabili.html. Accessed 15 Jan 2024

Cities Facing Climate Change

Domenico Passarelli

Abstract This research concerns the complex analysis of ecosystem relations in territorial areas and is aimed at identifying and developing sustainability indicators with the attempt to promote actions aimed at safeguarding and enhancing the natural, cultural and identity heritage of the territory. The ultimate goal is to build scenarios and predictive models for monitoring the impact of transformations. The inspiring principle is centered on the growing awareness that natural ecosystems must be maintained in dynamic balance to guarantee viable future growth through the provision of goods and services based on the logic of ecosystem services. The research work has as its objective the improvement of the quality of life and includes aspects relating to the environmental, social and cultural well-being of citizens linked to a better use of the categories of ecosystem services (life support, supply, regulation of cultural values). In particular, from an infra- and intra-generational ecological perspective, a strengthening of identity values is expected, of the usability of the natural and cultural heritage of smaller urban centers and internal areas, of the quality of green and blue infrastructures aimed at sustainable use of territorial and environmental resources. The expected results are measured on the basis of key stability indicators by the European Commission to monitor the diffusion of innovation at a regional level.

Keywords Vulnerability strategy · Policies · Involvement

D. Passarelli (✉)
Università degli Studi Mediterranea, 89124 Reggio Calabria, Italy
e-mail: domenico.passarelli@unirc.it

D. Marino and M. Monaca (eds.), *Generative Artificial Intelligence and Fifth Industrial Revolution*, Lecture Notes in Networks and Systems 880,
https://doi.org/10.1007/978-3-031-73880-7_14

1 Opportunities and Prospects for Urban Settlements

1.1 Strategic Directions and Regulatory Documents

Adoption of the Adaptation Strategy, adopted in 2013 by the European Commission, opened a concrete political perspective in addressing the challenges arising from climate change. The document "Elements for a National Strategy for Adaptation to Climate Change" identifies the main sectors by defining the strategic objectives and actions for mitigating impacts. This document identifies urban settlements as "the main culprits and main victims of climate change and, being predominantly artificial systems, they lack autonomous resilience and their ability to adapt will therefore be entrusted to conscious human action" [1]. It is a very important challenge that affects all institutional levels and in particular that relating to cities which have the duty to understand their vulnerability to climate change and the need to implement sustainable adaptive responses. In Europe there are many cities that have advanced significant experiences, Stockholm, London, Rotterdam, Stuttgart and Copenhagen. The latter is known for its ambitious sustainability agenda and the goal of becoming carbon neutral by 2025. The city is investing in infrastructure for sustainable mobility, promoting the use of bicycles, strengthening public transportco and implementing policies to reduce greenhouse gas emissions by focusing on reducing energy consumption and green mobility. San Kjeld residents were the first to experience a neighborhood adapted to climate change. An almost obligatory choice for Copenhagen which in 2011 was hit by a catastrophic cloudburst which caused around a billion euros in damage. Hence the decision for the port city to find a way to protect itself from the violence of the climate of the present and future [2]. But also in Italy there are interesting experiences such as the AMICA project (Adaptation and Mitigation Integrated Climate protection Approach; Interreg III, 2005–2007) which involved the city of Venice, as well as the Urban Heat Island—UHI project (2008–2011) aimed at prevention and reduction of risks associated with heat waves. Another interesting project is the GRaBS (Green and Blue Space Adaptation for Urban Areas and Eco Towns; Interreg IVC, 2007) which aimed to raise awareness among Administrations regarding the opportunity to increase natural or semi-natural networks within the city (Italian partners are the province of Genoa, the University of Catania and Etnambiente). Among the Regions, Emilia Romagna deserves to be highlighted as it already signed the "Under 2 Memorandum of Understanding" in 2015, which commits the organization to an 80% reduction in its emissions by 2050, defining the strategy for mitigation and adaptation. It was the beginning of a path of knowledge, integration and strengthening of regional policies for mitigation and adaptation, going well beyond what was requested by the European Commission with directives and regulations [3]. In 2013 the EU issued the Mayors Adapt as an initiative of the DG Climate Action of the European Commission included in the context of the EU Adaptation Strategy (http://bitly.com/EU adapt docs) with the aim of promoting the formation of Adaptation Plans urban areas, favoring the coordination role of the MATT, the Regions and the Provinces in order to allow not only larger urban entities, but also small municipalities, individual or associated, to equip themselves with this fundamental tool to combat climate change. In 2015, the "2030 Agenda for sustainable development" was signed, consisting of 17 Sustainable Development Goals (SDGs) framed in an action program divided into 169 targets or goals to be achieved by 2030 in each and all States under writers with reference to the different sectors of collective life (environmental, economic, social and institutional). A few years later, in 2019 in Brussels the EU began the path that in 2019 will take the name of "New Green Deal", and then in 2021 of "Next Generation-EU": It is precisely on the topic of urban settlements that, in the period between the SNAC and the PNACC, the Ministry of Ecological Transition (now the Ministry for the Environment and Energy Security) intended to promote one of the actions already envisaged by the Strategy, such as the "implementation of experimental climate adaptation interventions of public spaces in particularly vulnerable neighborhoods, increasing the greenery, the permeability of the soil, the social spaces, the hydraulic performances", launching, in collaboration with ANCI and ISPRA, the experimental program of interventions for the adaptation to climate changes in urban areas (MiTE, 2021). It is necessary to determine a set of integrated policies which, starting from the joint application of mitigation and adaptation policies, with an approach of strategic value, is capable of interacting and engaging the different levels of territorial management, intervention and above all the different subjects involved.

1.2 Urban Planning and Climate Change. Reverse the Current Trend

Today there is a need to implement a green strategy which can only start from the recognition of the important role played by greenery in urban settlements. The green deal favors the mitigation of seismic, hydrogeological and other environmental risks that limit and very often strongly inhibit the development of good planning practices. The new objective of urban climate agendas must therefore be twofold: the transformation must move both on the mitigation front, through the reduction of greenhouse gas emissions and the increase in the capacity to absorb CO_2, and on that of adaptation, aimed at making urban areas more resilient to climate threats (Ispra, 2023). There are many policies and tools that can be implemented starting from the new platform promoted by the EU following the pandemic for urban greening, drawing on the LIFE program which, among other things, recalled how urban green spaces (parks, gardens, green roofs and vegetable gardens) offered multiple benefits. Some measures, such as Agenda 21 and the Aalborg Charter, further highlight its extraordinary importance. Emilia Romagna with law n. 24/2017 relates climate change to the containment of land consumption, ecological and environmental facilities, the resilience of housing and urban systems (Art. 1, c. 2, letter a, Art. 21, c. 1, Art. 34 c. 1). The investigation of the climate framework and the phenomena that occur in the territory is aimed at identifying the consequent effects on urban planning, territorial security, people and liveability, in social and economic terms. We therefore reiterate the concept according to which with the increase in global temperatures and the ever-increasing incidence of extreme climate events, adopting planning strategies that integrate sustainable and resilient solutions becomes a fundamental step for urban and territorial management. The advantages of greenery are essentially ecological in nature as it is crucial for the improvement of the microclimate, for energy saving and for the sustainability of cities. Over the years it has increasingly taken on a social function without underestimating the fact that it increases citizens' safety and improves their mood, to the point of combating depression and loneliness. The Georgofili Academy in Florence presented some studies with surprising innovations. Trees are a valuable deterrent against petty crime, robberies and pickpocketing, and thus represent a safeguard for the safety of citizens. The European challenge for the transformation of cities into green and sustainable communities is corroborated by Mission 2—Green Revolution and ecological transition (divided into 4 Components: C1. Circular economy and sustainable agriculture; C2. Renewable energy, hydrogen, network and sustainable mobility; C3. Energy efficiency and redevelopment of buildings; C4. Protection of the territory and water resources) which sees the assignment of a budget of 69.93 billion euros, corresponding to 29.74% of the total amount of PNRR (of which 59.46 billion from the Recovery and Resilience Facility, 9.16 billion from the Complementary Fund and 1.31 from the React EU). In order to counteract the damage caused by climate change, the adoption of sustainable territorial planning can no longer be postponed by abandoning emergency action and moving towards an integrated strategy. With reference to urban settlements in par. 3.14 of the PNAC, intense precipitation events, extreme temperature events, increased periods of drought and rising sea levels are highlighted. Specifically, increases in global temperatures and alterations in precipitation patterns have led to an increase in intense precipitation events in urban areas, known as pluvial floods. These phenomena are characterized by precipitation events that exceed the absorption and drainage capacity of the soil and urban collection systems. The variability and magnitude of the phenomenon depend on the fact that cities are characterized by a high inter- and intra-specific heterogeneity, which contributes to accentuating the effects of pluvial flooding, and consequently not only topographic and morphological differences, but demographic and socioeconomic can also influence the severity of impacts. Addressing pluvial flooding requires integrated and coordinated action. It is necessary to adopt urban design approaches that take into account the management of rainwater (e.g. hydraulic invariance, lamination), promoting the creation of green spaces, soil permeability and the correct sizing of drainage systems. As mentioned throughout this writing, green infrastructures, conceived according to the valorization of nature and natural processes, must be integrated into territorial planning inspired by the DNSH principle (Do No Significant Harm), in sense that any activity in the area must not cause damage to the health of the inhabitants, the protection and restoration of biodiversity and ecosystems, the prevention and reduction of pollution, the circular economy. Some initial conclusive considerations reflect the need to evaluate the consequences of interventions on the territory through the application of multi-criteria evaluation techniques and in particular of the Strategic Environmental Assessment in the plan process, through the Integrated Assessment of Environmental and Health Impact (VIIAS), a combination of procedures, methods and tools (EIA—Environmental Impact Assessment, VAS—Strategic Environmental Assessment, VIS—Health Impact Assessment), with which the potential effects on health and distribution can be estimated of such effects within the population as part of current environmental assessment procedures. In addition to what has been expressed, the importance of guaranteeing the availability of resources aimed at making the ecological transition process effective in relation to the country's recovery and resilience plan (PNRR) and stopping the erosion of natural capital is underlined. The actions conceived in the course of this writing make explicit reference to the renaturalization of urban areas to mitigate the emissions of climate-altering gases and the health effects of fine particles, to encourage investments that reduce CO_2 emissions with a multimodality plan of new networks in order to support a plan to relaunch local transport systems with respect to high speed, cycle paths, pedestrian areas and activate the energy efficiency of both production plants and housing units with constraints that provide for at least three steps of category compared to the current two. At the same time, it would be appropriate to adopt measures to reduce and prevent waste, not only by intervening on recycling which, evidently, cannot be repeated for many types of goods.

1.3 Adaptive and Sustainable Tools. Some Guidelines Available to Cities

Remember that proposals for sectoral adaptation actions can be of different types: non-structural or "soft" actions; actions based on an ecosystem or "green" approach; infrastructural and technological or "grey" actions [4]. The actions that urban planning can undertake to address climate change must certainly be included in an ecosystem approach as part of urban development plans that promote energy efficiency, the reduction of greenhouse gas emissions and the use of renewable energy. NMP Some attention must be paid to sustainable construction, introducing regulations that promote the construction of buildings with low energy consumption or zero carbon emissions. Thinking of a new way of connecting and making the green spaces of our cities more accessible is an essential challenge! Law no. 10/2013 "Regulations for the development of urban spaces" did not produce obvious results. Cities can increase green areas and open spaces, creating parks and public gardens which also contribute to the absorption of carbon and the reduction of the heat island effect. One of the major problems is in fact the increase in CO_2 emissions into the atmosphere and therefore the increase in the average global temperature. The scientific evidence of the extent of global warming has become increasingly consolidated in recent years, as has the awareness that it is caused by the emissions of climate-altering gases deriving from the use of fossil fuels and the unsustainable use of land and natural resources. In order to counteract the ever-increasing impacts on human health, on sudden increases in energy demand, on shortages in drinking water supplies, on the drastic decrease in the quality of life of the most disadvantaged sections of the population, on the increase in risks associated to floods and soil instability and fires, "the Intergovernmental Panel on Climate Change (IPCC) calculates that polluting emissions must be reduced by around 45% compared to 2010 levels by 2030, reaching net zero by 2050" [5]. Especially in large settlements, some impacts are enhanced by the typically urban phenomenon of the Urban Heat Island (UHI), which induces higher temperatures on average in the central parts of cities compared to peripheral and extra-urban ones (up to 5–10° difference). Among the objectives that must be set is the development and application of indicators of effectiveness of the adaptation measures to be implemented. This means overcoming the predominantly quantitative approach and adopting a new standard based on a qualitative and performance criterion, such as the "ecological environmental quality standard" contemplated by the Regional Law of Emilia Romagna 20/2000, art. A-25 ecological and environmental facilities, paragraph 4, letter C, which indicates some specific measures and actions such as, for example, the provision of vegetated and tree-lined green spaces, basins and wetlands and the maintenance or creation of open spaces within the urban and peri-urban territory [6]. This means measuring, through physical, geometric and vegetation indicators, the performance of urban greenery (permeability, hedge lawns, shrubs, low-medium-tall trees) with respect to a series of functions: shading, permeability, production of oxygen O_2, absorption of carbon dioxide CO_2 and pollutants, acoustic insulation, creation of urban habitats and urban ecological micro-corridors to promote biodiversity, windbreak effect, division of spaces, visual isolation, etc. In addition to the well-known aesthetic and recreational functions, green areas contribute to mitigating pollution of the various environmental matrices (air, water, soil), improve the microclimate of cities and maintain biodiversity. With reference to the water resource, in the last thirty years its availability has decreased by 20%, according to the BIGBANG model of Ispra, compared to the reference value recorded in previous years [7]. There are many indicators developed at various levels by national and international bodies to pursue the objectives of urban sustainability and urban green spaces is one of these. The accessibility of public green areas and local services is also an indicator, measured by the percentage of the population living within 300 m of green areas of a size of at least 5000 m^2 (parks, gardens, open spaces, equipment, usable private green areas …) and some basic services (health, transport, education, food, greengrocers, etc.). The indicators that refer to urban greenery are: usable urban greenery: m^2/inhabitant of usable greenery, excluding parks and protected areas; green areas: surface area of the various green areas on the total municipal surface area (m^2/ha). Each requirement depends on a set of basic indicators which are measured via an index calculated as a weighted function of the basic indicators with respect to functionality, aesthetic quality, safety, services and furnishings, pressure factors. Furthermore, the quality of these areas is essentially measured according to their usability. Directly or indirectly, human beings depend on ecosystem services which, according to the definition proposed by the MEA, Millennium Ecosystem Assessment, are the "multiple benefits provided by ecosystems to mankind" (MEA, 2005) and can be grouped into 3 main categories: regulation SE, prevention SE, and supply SE. The theoretical model of "environmental services" is a very useful analysis tool because—by identifying in detail the various functions of nature and the multiple services it provides—it helps us to "translate" the benefits for society which "today more than ever, it is called upon to direct urban policies towards efficient management of the resources present in the city, giving greater attention to urban and environmental planning, to the creation of economically attractive places to live and functioning social spaces, to the construction of energy efficient systems" [8].

References

1. MATT, Ministero dell'Ambiente e della Tutela del Territorio e del mare (2014) Elementi per una strategia di adattamento ai cambiamenti climatici. Roma, p 104
2. https://ecobnb.it/blog/2023/06/copenaghen-ha-un-piano-zero-emissioni-dal-2025
3. Regione Emilia Romagna (2019) Direzione generale cura del territorio e dell'ambiente. Quaderno Cambiamenti Climatici. Documento di sintesi del 10.9.2019 work cited note 1
4. IPCC (2018) Global warming of 1.5 °C (an IPCC special report). In: 48th session of the IPCC. Incheon, Republic of Korea
5. Regione Emilia Romagna (a cura di), cambia il Clima, trategia per il cambiamento climatico
6. Carlo C (2023) La crisi climatica in Italia: un focus su Emilia Romagna. SBAM (Scuola di progettazione Bioclimatica per l'Adattamento e la Mitigazion), Webneir 01, 31 Mar 2023
7. Marino D, Trapasso R (2020) Servitization and territorial self reinforcing mechanisms: a new approach to regional competitiveness. Investig Reg—J Reg Res 48:85–93. https://doi.org/10.38191/iirr-jorr.20.022
8. Marino D, Gil Lafuente J, Tebala D (2023) Innovations and development of artificial intelligence in Europe: some empirical evidences. Eur J Manag Bus Econ 32(5):620–636. https://doi.org/10.1108/EJMBE-03-2023-0085

The manufacturer's authorised representative in the EU is Springer Nature Customer Service Centre GmbH, Europaplatz 3, 69115 Heidelberg, Germany. If you have any concerns regarding our products, please contact ProductSafety@springernature.com

Printed and bound by CPI Group (UK) Ltd, Croydon, CR0 4YY

30/09/2026

02225600-0002